Interoperability of DRM Systems

Forschungsergebnisse der Wirtschaftsuniversität Wien

Band 14

PETER LANG
Frankfurt am Main · Berlin · Bern · Bruxelles · New York · Oxford · Wien

Susanne Guth

Interoperability of DRM Systems

Exchanging and Processing
XML-based Rights Expressions

PETER LANG
Europäischer Verlag der Wissenschaften

Bibliographic Information published by the Deutsche Nationalbibliothek
The Deutsche Nationalbibliothek lists this publication in the Deutsche Nationalbibliografie; detailed bibliographic data is available in the internet at <http://www.d-nb.de>.

ISSN 1613-3056
ISBN 3-631-53845-6
US-ISBN 0-8204-7718-4

© Peter Lang GmbH
Europäischer Verlag der Wissenschaften
Frankfurt am Main 2006
All rights reserved.

All parts of this publication are protected by copyright. Any utilisation outside the strict limits of the copyright law, without the permission of the publisher, is forbidden and liable to prosecution. This applies in particular to reproductions, translations, microfilming, and storage and processing in electronic retrieval systems.

www.peterlang.de

Für meine Felsen in der Brandung: Papi, Mutti und Kerstin,

und für Christian, der mein Leben versüßt.

Acknowledgements

First of all, I would like to thank Prof. Gustaf Neumann and Prof. Alfred Taudes for supervising this thesis. In particular, I am indebted to Prof. Gustaf Neumann for numerous discussions and for his support in the rapid completion of this thesis. My thanks also go to my friends and all colleagues at the Department of Information Systems, especially to Mark Strembeck and Uwe Zdun who have always been open for discussions and who have given me a very agreeable working environment. Renato Iannella was always available for technical support and advice concerning the Open Digital Rights Language. I am indebted to Margit De Toma, who has accomplished most of the technical implementation of the rights expression generator. Further more I am grateful to my friend Tina Litschauer who spent a good deal of her spare time to correct my English vocabulary and grammar mistakes. Again, thanks to all of you without whose promotion this *promotion* would not have been possible.

Contents

Acronyms	13
List of Figures	14
List of Tables	16

1 Motivation   19
  1.1 Introduction 19
  1.2 The Impact of Standardised Contracts to Electronic Commerce   21
  1.3 Objectives of this Doctoral Thesis 23
  1.4 Classification into Research Theory 27
  1.5 Structure of this Doctoral Thesis 33

2 Digital Rights Management Systems   37
  2.1 Trading Digital Goods 38
    2.1.1 Characteristics of Digital Goods 38
    2.1.2 Business Models for Digital Goods 38
  2.2 Digital Rights Management (DRM) 41
    2.2.1 DRM Definition 41
    2.2.2 Perspectives of DRM 42
  2.3 A Sample Digital Rights Management System and its Functions   46
    2.3.1 DRM System Functions 47
    2.3.2 A Sample DRM System 53
    2.3.3 A Sample DRM Process 54
    2.3.4 Commercial DRM Products and
           DRM System Variants 59
  2.4 The Role of Rights Expression Languages in DRM 63

3 Rights Expression Languages (RELs) — 67
- 3.1 Definition of Terms 67
- 3.2 Requirements of RELs 68
- 3.3 Characteristics of RELs 70
  - 3.3.1 REL Syntax 71
  - 3.3.2 Rights Data Dictionary (RDD) 72
- 3.4 Existing Rights Expression Languages and Initiatives 73
  - 3.4.1 Open Digital Rights Language (ODRL) 73
  - 3.4.2 eXtensible rights Markup Language (XrML) 81
  - 3.4.3 MPEG 21 83
  - 3.4.4 LicenseScript 84
- 3.5 Current Market Situation and Trends 84

4 Electronic Contracts — 87
- 4.1 Contract Life Cycle 89
- 4.2 Contract States 90
- 4.3 Execution of Rights 92
  - 4.3.1 Electronic Contracts, Electronic Tickets, and Licenses 93
  - 4.3.2 Ticket-Driven Rights Execution 95
  - 4.3.3 Hybrid Rights Execution 97
- 4.4 Contract Objects and Contract Use 98
  - 4.4.1 Core Contract Objects 98
  - 4.4.2 Sample Usage Scenarios for Electronic Contracts . . . 101
  - 4.4.3 Scenario-Specific Contract Objects 103
- 4.5 Contract Modelling and Creation 104
  - 4.5.1 Required Information for Specific Software Services . . 105
  - 4.5.2 Modelling Scenario-Specific Contracts 109
  - 4.5.3 Scenario-Specific Contract Composition 112
- 4.6 The Generic Contract Schema 115
  - 4.6.1 Definition of Terms 115
  - 4.6.2 Application-Specific CoSa Example 117
  - 4.6.3 The CoSa API 121
  - 4.6.4 CoSa Serialisation 124
- 4.7 Enforceability of Electronic Contracts 126
- 4.8 Contract Management Issues 129
- 4.9 Related Work 138

5 Design of a Rights Expression Exchange Framework 145
- 5.1 Exchanging Rights Expressions 145
  - 5.1.1 The Communication Model 146
  - 5.1.2 The Rights Expression Communication Model 147
- 5.2 The Rights Expression Exchange Framework 149
  - 5.2.1 Technical Design 150
  - 5.2.2 Implementation Check List 153

6 Implementing the Rights Expression Exchange Framework 157
- 6.1 Software Architecture 157
  - 6.1.1 The XOTcl Language 158
  - 6.1.2 ActiWeb 160
  - 6.1.3 Document Object Model (DOM) Implementation ... 161
  - 6.1.4 MySQL 163
  - 6.1.5 OpenSSL 163
- 6.2 The Rights Expression Generator 164
  - 6.2.1 Functional Description 165
  - 6.2.2 Class Hierarchy 168
- 6.3 The Rights Expression Interpreter 169
  - 6.3.1 Functional Description 170
  - 6.3.2 xoREL Packages and Classes 172
  - 6.3.3 Mapping ODRL Elements to the Contract Schema .. 176
- 6.4 The Rights Expression Wrapper and Unwrapper 179
  - 6.4.1 Functional Description 180
  - 6.4.2 Class Hierarchy and API 181
- 6.5 The Mediator 182
- 6.6 Implementation Assumptions 183
- 6.7 Related Work 184

7 Case Study of the Rights Expression Exchange Framework 189
- 7.1 Access Control with Context Constraints 191
- 7.2 Access Control Decision Based on Electronic Tickets 196
  - 7.2.1 Application–Specific CoSa 197
  - 7.2.2 Generating DRM–Specific Licenses 198
  - 7.2.3 Wrapping DRM Licenses 200
  - 7.2.4 Unwrapping, Interpreting and Processing DRM Licenses 200

8 Conclusion and Future Work 209

9 Appendix A 215
9.1 ODRL Foundation Model 215
9.2 XML Schema of ODRL Syntax Version 1.1 215
9.3 XML Schema of ODRL Data Dictionary Version 1.1 222

10 Appendix B 229
10.1 CoSa Application Programming Interface 229
10.2 Extended CoSa Application Programming Interface 234
10.3 Wrapper / Unwrapper Application Programming Interface . . 237

Bibliography 241

Index 260

Acronyms

API	Application Programming Interface
ATM	Automated Teller Machine
B2B	Business to Business
B2C	Business to Consumer
C2C	Consumer to Consumer
CoSa	Contract Schema
CRM	Customer Relationship Management
DAC	Discretionary Access Control
DOI	Digital Object Identifier
DOM	Document Object Model
DTD	Document Type Definition
EDI	Electronic Data Interchange
FTP	File Transfer Protocol
HTML	Hytertext Markup Language
HTTP	Hypertext Transfer Protocol
IEC	International Electrotechnical Commission
IPR	Intellectual Property Rights
IS	Information Systems
ISBN	International Standard Book Number
ISO	International Organization for Standardization
ISSN	International Standard Serial Number
LOM	Learning Object Metadata
MAC	Mandatory Access Control
MIS	Management Information Systems
MPEG	Moving Picture Experts Group
OCR	Optical Character Recognition
ODRL	Open Digital Rights Language
OMA	Open Mobile Alliance
PDA	Personal Digital Assistant

PDF	Portable Document Format
PHP	Hypertext Preprocessor
PKI	Public Key Infrastructure
RBAC	Role Based Access Service
RDBMS	Relational Database Management System
RDD	Rights Data Dictionary
RDF	Resource Description Framework
RE	Rights Expression
REL	Rights Expression Language
SGML	Standard Generalized Markup Language
SIM	Subscriber Identity Module
SQL	Structured Query Language
SSL	Secure Socket Layer
Tcl	Tool command language
TCP	Transmission Control Protocol
TLS	Transport Layer Security
W3C	World Wide Web Consortium
XML	eXtensible Markup Language
XOTcl	eXtended Object Tcl
XrML	eXtensible rights Markup Language

List of Figures

2.1	The six perspectives of DRM	42
2.2	The DRM perspectives in the order of their influence on DRM systems	45
2.3	Basic and extended functions of DRM systems	48
2.4	A sample DRM system	53
2.5	A sample DRM process	56
2.6	InterTrust's DRM system	61
3.1	A subset of the ODRL language syntax	75
3.2	A simplified subset of XML schema defining ODRL	77
3.3	A valid language instance of the simplified ODRL schema	78
4.1	A simple contract life cycle with four phases	89
4.2	Basic states and state transitions of electronic contracts	91
4.3	*Contract right* versus *permissions*	93
4.4	Contracts and tickets – an example	94
4.5	Combination of tickets and direct rights processing	97
4.6	The abstract core objects of electronic contracts	99
4.7	Various usage scenarios for electronic contracts	104
4.8	Assigning permissions in RBAC	106
4.9	Application–specific data model	110
4.10	Example of mapping of objects instances and their attributes to software services	112
4.11	Composing tailored electronic contracts	114
4.12	Class diagram of an application–specific Contract Schema	118
4.13	Application–specific contract schema	119
4.14	Instance of an application–specific Contract Schema	120
4.15	The enforceability matrix	128

4.16 Sample operations when managing electronic contracts 130
4.17 General structure of an service level agreement [LDF+02] . . 139
4.18 Simplified model of contracts applied in a WFWM [KGV99] . 141

5.1 The communication model [Sch71] 146
5.2 The rights expression communication model 147
5.3 Components of a rights expression exchange framework . . . 151

6.1 Technology used in the rights expression exchange framework 158
6.2 Features of XOTcl, OTcl, and Tcl 159
6.3 Basic architecture of ActiWeb [NZ00a] 160
6.4 A general DOM–tree 162
6.5 Choice of ODRL tags 166
6.6 Display and store generated ODRL rights expression 166
6.7 Choosing constraints via the customised generator GUI . . . 167
6.8 Reused software packages in the rights expression generator . 168
6.9 Class hierarchy of ODRL specific elements 169
6.10 Functional layers of XML–based rights expressions 170
6.11 The interpretation process 171
6.12 Classes of the package `contract` 173
6.13 Classes of the package `reInterpreter` 175
6.14 Packages with wrapping respectively unwrapping functionality 182
6.15 The mediator, using framework components and other packages 183

7.1 Execution of an access request 193
7.2 Sample access permission with constraints 193
7.3 xoRBAC access control decisions with context constraints . . 195
7.4 The application–specific CoSa 198
7.5 Provide license templates with the generator 199
7.6 Mediator code combining generator and wrapper functionality 201
7.7 Sequence diagram with basic activities of the secure viewer . 202
7.8 Runtime model of the DRM CoSa objects 204

9.1 The foundation model of ODRL [Ian02b] 216

List of Tables

4.1 Characteristics of application–specific and domain–specific CoSa 117
4.2 Possible role names in application–specific CoSa 121

6.1 Possible role names in application–specific CoSa 174
6.2 Mapping of ODRL asset context to CoSaResource objects . . 178
6.3 Mapping of ODRL party context to CoSaParty objects 179
6.4 Mapping of ODRL agreement/offer context to CoSaContract objects 180

Chapter 1
Motivation

The first section shall deliver an overall insight about this doctoral thesis. Section 1.1 gives a short introduction to the exchange of rights on goods or services within the scope of electronic commerce (short: e–commerce) and gives some examples of drawbacks that the current non–standardised e–commerce technology has. Section 1.2 then provides the theoretical background why standardised technology can improve and quicken e–commerce. In Section 1.3 the objectives of this doctoral thesis are defined. The subsequent section deals with the classification of this thesis into research theory (see Section 1.4) and introduce the research methods that have been applied. The first chapter closes with an overview of the overall structure of this thesis (see Section 1.5).

1.1 Introduction

The Internet has evolved as worldwide e–commerce platform for digital goods. Technically, digital goods are any kind of information that can be digitised, such as baseball scores, books, databases, software, magazines, movies, music, stock quotes, and web pages [SV99]. Most of the existing digital goods did not evolve with the penetration of the Internet, but simply changed their medium from physical to digital. The digital medium offers a large spectrum of new ways for commercialisation [WIP02]. Ten years ago, for example, one obtained a certain song by buying the complete album for €10.00 in form of a compact disc or on vinyl, i.e. there was only one way to purchase this song, because it was bound to the medium of a compact disc. Producing a compact disc for each song on the album is far too cost intensive and risky. By using the Internet as medium, these boundaries melt and it is possible to bring this song to market in various forms: as a single song for e.g. €1.00, in a bundle of 5 songs for €3.00, and with the restriction to play the song only five times for €0.50. The production costs of these three products are low, as the copying and bundling of digital goods is cheap.

The Internet gains increasing importance as distribution channel for digital goods [Bak98, Zhu01]. The distribution via the Internet is also cheap for the seller, as normally the consumer pays the carrier costs. People and companies from all over the world have business with each other trading both physical and digital goods. The basis for each business is a contract. Today's electronic business works with *electronic* contracts concluded between two dislocated contracting parties that are not very transparent. An *electronic contract* is an agreement of two or more parties, on the exchange of rights to (digital) goods or services under certain terms and conditions. The memorandum of an electronic contract is digital and can be transmitted via an electronic network (see Section 4).

With a simple click, you can purchase a book, a song, a washing machine. What is missing, most times, is the contract information in a format that enables consumers, vendors, and third parties to reconstruct what exact product has been purchased, who the contract parties are, and which permissions and duties have been agreed on. Nowadays, contracts that are concluded via the Internet are either not explicitly stored or kept as database entries in a proprietary format, containing relevant information for the processing software system. For example, the EducaNext platform[1], a brokerage platform for learning resources, stores booking id, resource id, user id, and optionally a comment, after a learning resource has been booked. This fact has a number of drawbacks:

- People enter into proprietary contracts with each business partner. The non–standardised representation of terms and conditions is not transparent or comparable.

- The contracts can not be presented to or verified by a third person (e.g. a lawyer or a bank) because in most cases electronic contracts are ambiguous and/or not readable for humans.

- The contracts can not be exchanged between platforms that do not operate with the same software or did not agree on a common contract format.

---

[1]See: http://www.educanext.org/

- Because of the non–standardised format of contracts, most times electronic contracts have to be fulfilled with the platform where they have been concluded. For example, one can not buy a contingent of 100 music units and spend it at arbitrary online music providers.

- Contracts also contain valuable information for various business applications (access control, book keeping, customer relationship management, etc.). Today, most contracts are designed to be processed in a single application. Changing the contract structure or the contract content usually requires an adaptation of database tables and software.

- In the pricing model (see Section 2.1) 'playing a certain music file 5 times for €0.50' the contract comprises the constraint information (play only five times). This information usually has to be interpreted by the digital player (also called secure viewer. To understand and enforce the constraint information, the player has to understand the proprietary format. As a result, today's online music shops all have their own player, to the displeasure of the customer.

Most likely, this is not a complete list of disadvantages, but it shows a number of severe drawbacks when using proprietary contracts. This thesis aims at giving a basis to reduce these disadvantages. To reason the objectives of this thesis more profoundly, the subsequent chapter describes the theoretical background to the advantages of deploying standardised electronic contracts.

1.2 The Impact of Standardised Contracts to Electronic Commerce

Electronic Commerce is the 'business activities conducted using electronic data transmission via the Internet and the world wide web' [SP00]. Several parties profit from e–commerce, e.g. consumers and providers of physical and digital goods: consumers benefit from accessing global markets for electronic and physical goods via the Internet. In turn, content providers can sell their products and services on global markets and in narrow market segments that may be geographically distributed [NJRW01]. Apart from providers and consumers, many additional actors will profit from e–commerce [Rig03], namely the actors in the business to business (B2B) context that are involved in the supply chain: e.g. actors in procurement, content creation, content packaging, content publishing, content selling, content distribution, content consumption services, and customer relation ship management [Sup03, Ian03b, Sun02, Ass00].

Cooperation among various actors in the supply chain results in a network of actors on the Internet. The term 'Metcalfe's Law'[2], sometimes also called *network effect*, that was coined by Gilder [Gil93], states that the value of a network grows with the square of the number of participants. In other words, each additional member of a network adds an incremental amount of value to every other member, thus increasing the aggregate value of the network in a quadratic fashion, while the cost–per–user remains the same or even decreases. This means that actors cooperating in the supply chain build an e–commerce network that is more valuable than a number of isolated actors. For example, a number of content publishers found a consortium to offer their goods via a common portal and agree on using a certain payment method. If their customers trust this consortium and find the payment method comfortable and secure, a new member of the consortium will easily be accepted and trusted by the customer as well. The consortium reaches more customers by affiliating new members, and the customer has more products at the favorite content publisher portal to choose from.

In practice, organisations have aspired to achieve better collaboration with partners via interorganisational systems like electronic data interchange (EDI) that permit firms to exchange electronic information. However, most attempts have been focused on pairs of business partners rather than on providing an open standardised solution [CM03]. Such technology is hindering the network effect because a data interchange with all partners in a large community is not intended. Besides earlier attempts of cooperation have not been based on legally binding contracts which has caused increasing costs for managing collaboration and reduced its advantages [WBK03]. Accordingly, a general solution is required that facilitates the standardised data interchange of business to consumer (B2C), business to business (B2B), and also consumer to consumer (C2C). In particular, standardised contracts are required to state legally binding obligations between providers and sellers of goods and services in an electronic market [GSSS00], i.e. all actors in the supply chain.

To underline the importance of such standardised technology, the subsequent paragraphs will present studies that have investigated the perceived

---

[2] Attributed to Robert Metcalfe, originator of Ethernet and founder of 3COM

strategic value of information technology (IT) in general, the strategic value of e-commerce systems in particular, and the influencing factors to adopt e-commerce systems.

- *Strategic value.* In most studies the perceived strategic value of information technology focuses on the relationship between IT and the firm's performance. In [HE96] is was found that IT increases productivity and consumer surplus but not necessarily business profits. In [GP03] it is therefore concluded that "IT investments are important to maintain competitive parity but do not necessarily support competitive advantage". For e-commerce, as one particular field of a company's IT, [SS02] found that the most important area in which e-commerce will create value is in reducing transaction costs involved in bringing consumers and suppliers together. After reviewing a number of relevant studies in this field [GP03] identifies *organisational support, managerial productivity,* and *decision aid* as strategic reasons for e-commerce.

- *Technology adoption.* Statements on information technology adoption are often based on the Technology Acceptance Model (TAM) developed by Davis [Dav89]. In [LMSZ00] investigations based on TAM on different applications, such as e-mail, Internet, ATM, and MS Word, have been summarised. The results show that *ease of use* and *perceived usefulness* are the major factors that affect the intentions to technology use. *Perceived benefits, organisational readiness,* and *external pressure* were considered to be important factors for the adoption of EDI technology in the study of [IBD95]. In [GP03] a study with small and mediums sized businesses identified that the factor of *compatibility* is highly influencing the adoption of e-commerce technology.

1.3 Objectives of this Doctoral Thesis

Standardised electronic contracts are an important building block for compatibility and consequently a basis for cooperation between companies and customers doing electronic commerce. Studies about the adoption of electronic commerce clearly show that among other factors the perceived ease of use, the perceived usefulness, and especially the compatibility of e-commerce systems are the main reasons for the adoption of e-commerce systems (see Section 1.2). Therefore, electronic contracts as an elementary constituent of electronic commerce need to be easily exchangeable and processable to provide compatibility for e–commerce systems. Thus, electronic contracts have the potential to acquire the same acceptance, validity and trustworthiness as contracts in the paper world, and to quicken electronic commerce. To achieve this goal technically, a number of issues have to be solved: standardised *syntax, semantics*, and *processing approach*. Electronic contracts must have standardised syntax and semantics to be meaningful, machine readable, human and machine interpretable and revisable. Standardised syntax and semantics provide a language that, in turn, allows for exchanging contract information. Additionally, a standardised way of 'using' electronic contracts in software services is required. Here, software services are running programs that process electronic contracts such as access control services or accounting software.

- *Standardised Representation (Syntax) of Electronic Contracts.* To make contracts easily exchangeable, first of all, a standardised representation of electronic contracts is needed. One alternative is to linearise the contract content with the help of the eXtensible Markup Language (XML) [BPSMM00]. XML provides a framework to define the syntax of electronic documents, i.e. the structure and the allowed character set that may occur in documents. The structure and the allowed character set of an electronic contract is then defined in an XML schema [TBMM01, BM01] or XML Document Type Definition (DTD) [BPSMM00]. The nature of XML–based documents provides machine readability. In fact, the machine readability comes with XML parsers that understand the XML schema or DTD and accordingly read XML documents. Today, a large number of commercial and non–commercial XML parsers as well as XML document creators are available.

- *Standardised Semantics of Electronic Contracts.* To be human and machine interpretable electronic contracts require clear semantics. Contract semantics are defined, e.g. in the specifications of *rights expression languages* (see Chapter 3). Rights expression languages are XML–based languages capable of expressing rights of parties over assets. An instance of a rights expression language is a rights expression. Depending on their content, rights expressions can represent different semantic constructs, e.g. licenses, digital tickets, or contracts. Being machine *readable* simply means that a parser is able to extract the XML-tag names and their values from electronic contracts. Machine *interpretable* denotes a semantic analysis of the tags and values of the electronic contract. For example, the XML parser reads an XML–

tag with the name *party* and the value *Department of Information Systems* at a certain location in the XML contract document. The parser is not able to determine whether this party is the consumer, the provider, or simply a beneficiary. To assign the correct meaning to this information an interpreter that is familiar with the specification of the respective rights expression language is required. A language specification, in this context, is a text document or formal semantics that defines the semantics for the elements in a specific XML–schema or DTD. As an alternative to XML, the resource description framework (RDF) [LS99, BG00] could be used to define contract semantics. In contrast to XML, RDF is independent of a specific linearisation technique.

- *Standardised Processing of Electronic Contracts.* Clear syntax and semantics, respectively machine readability and interpretability are prerequisites for processing electronic contracts. Partly the pragmatics, or more exactly the processing task of electronic contracts has to be standardised as well, thus ensuring a common way of *using* electronic contracts in software services. Examples for processing electronic contracts are:

    - *Enforcing Electronic Contracts.* Enforcing electronic contracts is one way of *processing* them. Enforcing is the act of implementing access rights as stated in electronic contracts. Electronic contracts are technically enforceable if: means for processing are provided, the contract comprises all required information, and a 'secure' *enforcement software service* (e.g. an access control service) is available.
    - *Human Readability of Electronic Contracts.* To be human readable and additionally human revisable the contract information has to be represented in a sensible layout, without showing the XML–typical tags or tree structure. To provide human readability, a software service has to be available that arranges the contract information in a sensible layout.

    In the two above examples and most likely in all other contract usages it is important that two dislocated, independent software services have the same effect (or result) when processing an identical electronic contract.

In addition to the above mentioned technical challenges, the handling of electronic contracts additionally has legal, managerial, and security challenges. For example, it is important to decide whether a contract is legally and technically valid or not, or to uniquely identify contract parties and objects. Besides, it is crucial to know which contract content is mandatory for a specific software service (such as access control service or accounting services) and if, or to what extent the contract content is electronically enforceable.

> This work aims at developing methods[3] and tools for exchanging and processing XML–based rights expressions (in particular electronic contracts) in consideration of legal, managerial, and security issues.

The methods and tools (see Section 1.4) aim at reaching the goal, i.e. compatibility via standardised representation, semantics, and processing, and thus address the drawbacks mentioned in the introduction. We are not aware of any methods and tools that are currently available for exchanging and processing of electronic contracts (generally speaking: rights expressions). The methods to be developed need to be of generic nature and independent of a specific technology (e.g. the programming language or the used rights expression language). The tools should be available at least as prototype implementations that are open and extensible and prove the correctness and usability of the introduced methods.

More precisely, my work is focused on designing and implementing a rights expression exchange framework that facilitates the encoding, transmission, and decoding of rights expressions for subsequent processing in software services. With regard to standardisation, the rights expressions shall be formulated in a rights expression language (see Chapter 3). The components require a well defined interface to assure its (re)use in various environments and to ease its integration into existing systems. The implementation shall be coded in an appropriate programming language, reuse existing technology, and consider relevant standards. The deployment of the resulting software components shall be demonstrated in a concrete use case.

Standardisation of processes and interfaces can be used to achieve compatibility [KS94, FS92]. According to studies mentioned in Section 1.2,

---

[3]A method can be understood as a procedure or concept that is comprising steps to reach a certain goal.

compatibility is a driving factor for the adoption of e-commerce systems. Therefore, I conclude that this thesis has the potential to leverage the adoption of e-commerce technology. E-commerce platforms that facilitate the distribution of digital content via standardised, electronically processable rights expressions additionally have potential to reduce distribution costs and thus can have strategic value [Rig03]. However, these platforms should have in mind to find the right balance (described in [Les01]) between copyright protection and freely available digital content.

1.4 Classification into Research Theory

Firstly, this section defines the term 'Wirtschaftsinformatik' and proposes its translation into English. Then the general classification of the discipline 'Wirtschaftsinformatik' into existing sciences is sketched. Sciences can be classified depending on their *research objects*, their *research goals* and their *research methods*. These three criteria will be used to classify the discipline 'Wirtschaftsinformatik' in general and the work at hand in particular.

This thesis is handed in to acquire the doctoral grade in the field of 'Wirtschaftsinformatik'. 'Wirtschaftsinformatik' is a German term that has no common translation into English. It is sometimes referred to as "Business Information Systems (BIS) Science", "Management Information Systems" (MIS), "Business Informatics", or simply "Information Systems". There is an ongoing debate in the German speaking countries about an appropriate English term. In a panel discussion of the German professors Buhl, Mertens, Koenig, and Krcmar it was stated that the German translation for *Information Systems* is 'Wirtschaftsinformatik' [BKKM97]. Despite this definition, in this thesis the term *business information systems* (short: BIS) will be used as the translation of 'Wirtschaftsinformatik', representing the field of information systems that has its focal point on business issues and often applies constructive research methods (see Section 1.4). However, a common agreement exists that 'Wirtschaftsinformatik' has an interdisciplinary orientation between the fields of economics, business administration, and computer science [BKKM97, KHvP95, Fra99].

In general, science is divided into *formal science* and *informal science*. In formal sciences, such as mathematics and logic, formal languages are developed that do not have a relation to real objects. Informal sciences deal with the *description, explanation,* and *design* of empiric objects. Informal sciences can be divided into fundamental science and applied science. Whereas fundamental sciences have the goal to *describe* and *explain* empiric objects, applied sciences have the goal to investigate the *design* of sociotechnical systems.

As mentioned above, BIS combines the fields of economics, business administration, and computer science. Computer science is an applied science, whereas economics and business administration can be both, fundamental science and applied science. Therefore, BIS is classified as an informal science, comprising both fundamental and applied sciences. Because of the interdisciplinary orientation between computer science, economics, and business science and the relatively young history, business information systems is lacking a common vision about research object and goal. This difficulty has been mentioned and criticised by various researchers [KHvP95, BKKM97, Fra99, MH02]. In the subsequent paragraphs visions from different researchers about research object and goal of BIS are introduced with the ambition to give a general overview of this matter.

Research goals of business information systems

In 1993, the German research commission of information systems defined the research goals of BIS as follows:

> The objective of business information systems is to gain *theories, methods, tools and reviewable knowledge* about/to information- and communication systems and to add methods and tools of business information systems that customise the sociotechnical knowledge and composition subject of scientific studies, to the "scientific case of methods and tools".[Wis94]

Research objects of business information systems

Two years later, in a panel discussion [KKK+95] the German professors H. Krcmar, W. König, K. Kurbel, D. B. Pressmar, A–W. Scheer, and W. Stucky named the following issues as current research objects of BIS:

- *Distributed information systems in business and management.* W. König: "The long term research goal in this field is to develop a normative co–ordination scheme for interacting actors. We aim to better integrate methods, tools, and applications, thus connecting human actors and machines more economically ... "

- *Empirical research on information systems in Germany.* H. Krcmar: "This research comprises survey methods, case studies, ethnographic studies, and laboratory studies in the field of BIS".

- *Parallel Processing and business applications.* D.B. Pressmar: "The application of evolving technologies to optimisation problems in production planning, such as scheduling and lot sising. Computational power is also needed when building neural networks to problems of pattern recognition such as forecasting or diagnosis in quality management and controlling".

- *Research on petri nets.* K. Kurbel: "This research field is especially focused on process modelling. Petri Nets are proposed as formal description language for business processes".

- *The influence of [BIS] research on industry.* A.–W. Scheer: " The BIS research is strongly dedicated to the development of prototypes. The ideas behind the prototypes should be included into commercial software products and improve applied information systems".

In the same year, the findings of an investigation, that tried to identify the essential **research objects** and theories of BIS with the help of the Delphi and AHP (Analytic Hierarch Process) method were published [KHvP95]. According to this investigation the four most important research objects in BIS are:

1. *Science with a strong relation to organisation theory.* This approach tries to describe and optimise the structure and process flow of sociotechnical systems.

2. *Functional business administration.* This discipline is investigating particularly the role of data processing and information processing in companies.

3. *Information science.* Information science aims at exploring the economy of the power factor "information" and its purposive allocation.

4. *Innovation science.* This discipline defines requirements for new information and communication techniques, and implements the resulting products and processes in companies.

Research goals of this thesis:

The research goal of this work is to develop methods, tools, and reviewable knowledge in the field of BIS, namely:

- *methods*, such as, the mapping of electronic contracts (respectively rights expressions) to a generic contract schema (see Section 4.6), or the process for composing tailored contracts (see Section 4.5),

- *tools*, e.g. the prototype implementation of a software tool that facilitates the generation of rights expressions, or the prototype implementation of a software tool that facilitates the interpretation of rights expressions, or (see Chapter 6), and

- *reviewable knowledge* e.g. about typical functions in DRM systems, the core components and the application–specific components of electronic contracts, characteristics of rights expressions languages, the constituents of a rights expression exchange framework, etc.

Thus this thesis meets the requirements in accordance with the research goal of the German research commission of information systems [Wis94].

Research objects of this thesis:

In general, the research object of this thesis is to provide technology to facilitate economic concepts. The methods and tools developed in this work shall provide technical means to support new business models for digital goods, to support technical means to meet the legal requirements of electronic contracts (e.g. providing digital signature information), as well as to provide new functionality to DRM systems in general, such as interoperability. Therefore, the thesis at hand models the processes of generating, wrapping, transmitting, unwrapping, and interpreting electronic contracts in a generalised way. In particular it addresses the processing of rights expressions in subsequent software services. As rights expression languages play a substantial role in this field, an empiric survey on rights expression languages is necessary. The work shall present the design and implementation of software tools that are capable of generating, wrapping, unwrapping, interpreting, and processing rights expressions. The work shall include a prototype implementation of such a general design and integrate the prototypes into different software environments. The overall focus of this work is to better integrate human actors and machines by providing means for the comfortable handling of electronic contracts. This thesis addresses the essential research objects (integration of applications, process modelling, surveys, and prototype development) of BIS in accordance with [KKK+95].

Research methods of business information systems

In [LHM95] it is stated that: ".. [BIS] applies methods and tools out of formal, informal and engineering science and develops them. The sociotechnical cognition subject of [BIS] demands that not only questions of technical efficiency but also questions of economical and social utilisation (including the acceptance of different social groups) are considered within the choice and combination of the used methods and tools."

In 1997 H.U. Buhl et al. three dominant methodological orientations in business information systems are named [BKKM97]:

1. *Engineering.* This approach aims at developing and testing software prototypes, including the design and application of formal modelling methods.

2. *System integration.* This approach focuses on organisational aspects of introducing and using information systems.

3. *General Models.* This approach aims at the development and analysis of formal models in order to support optimisation and decision making in general.

Most researchers feel committed to one of these three approaches exclusively. The combination of the various disciplines in BIS result in different methodological research orientations, depending on the weight of each discipline in the research work.

Empirical methods as well as *constructive* methods are the basic research methods in the field of BIS [Hol99, KHvP95]. The BIS incorporates the anglo–american MIS research as well as application oriented aspects of computer science; thus, compared to MIS research its methodologies are more constructive [ASBA99].

- Empirical research follows the process of problem analysis, i.e. developing a theoretical model and testing the model, data analysis, and interpretation. The interpretation of the results of such an analysis produces new knowledge.

- Constructive methods follow the idea that new knowledge is gained by constructing new ideas or concepts based upon the researcher's knowledge. Accordingly, when deploying constructive methods, the researcher does not try to verify predefined theories.

The discussion in the previous paragraphs have illustrated the disaccord on the objects of research in BIS. This disaccord causes some researchers to question if BIS actually needs its own research methods and theories [KHvP95, LHM95]. In practice, researchers avail themselves of methodologies from the related "mother" sciences business science and computer science.

Research methods applied in this thesis

In this thesis empirical methods as well as constructive research methods are applied. In general, this thesis follows the process of problem analysis, comprising the phases *problem analysis, development of theoretical model, testing the model*, and *analysis and interpretation*. The empiric objects in this process are electronic contracts, or, more generally, electronic (or digital) rights expressions:

- *Problem analysis*. Rights expression languages are a means to express usage and access rights of parties over (digital) assets. Such languages support the trading of electronic goods via the Internet. There are different initiatives that are developing language syntax and vocabulary for rights expression languages, but no general concept has been discussed or designed that describes the exchange and processing of rights expressions. The problem and research question of this work is: "What does a framework design that is capable of exchanging and processing rights expressions to support DRM system interoperability look like?"

- *Development of a theoretical model*. This phase comprises the design development of a framework that is capable of *exchanging and processing* rights expressions (e.g. in the form of electronic contracts).

The *theoretical model for the exchange of rights expressions* signifies that the communication model [Sch71] of Schramm illustrated in Figure 5.1 (a further development of Shannon's basic communication model [Sha48]) can be adapted to the communication via rights expressions. The resulting *rights expression communication model* is the basis for the rights expression exchange framework, consisting of the four tools: rights expression generator, – wrapper, – unwrapper, and – interpreter.

The *theoretical model for rights expression processing* is the *generic contract schema* (see Section 4.6). It describes a generic data model that serves as a basis for rights expression processing and facilitates an approach of the semantically unambiguous representation and processing of electronic contracts.

- *Testing the model (Implementation).* The implementation of a prototype is a means to test the previously developed framework design. The framework implementation naturally incorporates the theoretical model of rights expression processing (i.e. the generic contract schema).

- *Analysis and Interpretation.* The integration into various system environments allows an analysis of the framework design and its usability. A concrete application shows the usability for a certain domain (e.g. processing contract in the educational domain). The interpretation of the analysis' results are part of the findings of the work at hand.

Constructive methods have been applied during the implementation phase. For example, object oriented concepts have been used to structure and develop the tool for exchanging and processing rights expressions. The process definition for scenario–specific contract composition (see Section 4.5.3) has been developed in a constructive manner.

1.5 Structure of this Doctoral Thesis

The work at hand has the following further structure:

- Chapter 2 is an introduction to the area of digital rights management systems. Firstly, the chapter deals with the general commercialisation of digital goods. Consequently, it addresses characteristics, dimensions, and business models for digital goods. The chapter discusses various definitions of the term *Digital Rights Management* (DRM) and introduces the perspectives of DRM. The basic functions of a DRM system are defined, namely: *Content Provision, Content Safekeeping, Offer Placement, Content Preparation, Content Distribution,*

Booking, Payment, Authorisation, and Content Consumption. A subsequent section describes a sample system comprising all basic DRM functions that are grouped into typical system components. The information flow through such a sample system, which is basically the interaction between modules that implement a certain function and components, is described in detail. Finally, the role that rights expression languages play in DRM systems is explained.

- Chapter 3 gives an introduction to the field of rights expression languages. The chapter addresses the requirements and characteristics of rights expression languages. It introduces relevant existing rights expression languages, gives some practical examples, and shows their deployment in current non–commercial and commercial software products.

- Chapter 4 deals with relevant aspects of handling electronic contracts. Electronic contracts can be expressed in rights expression languages. Thus, the contracts have a standardised representation that facilitates the exchange of contracts between interoperating platforms. The chapter introduces a contract's life cycle and discusses typical technical states. It further presents means for rights execution, i.e. the technical fulfillment of contracts. The contract content is analysed in detail; core elements as well as application–specific elements of contracts are identified. The chapter discusses potential exploitation (or usage scenarios) of electronic contracts and describes the process of creating *tailored* electronic contracts. The generic contract schema (CoSa) is introduced; the generic CoSa is a concept for a (rights expression) language independent representation of electronic contracts. The chapter finally addresses the enforceability of electronic contracts, management issues when handling electronic contracts, and relevant related work.

- In Chapter 5 a general communication model facilitating the exchange of rights expressions is introduced. From the model a software framework, namely the rights expression exchange framework, is derived. This framework consists of the software components *rights expression generator, –wrapper, –unwrapper,* and *interpreter.* Additional sections describe the detailed functionality of all framework components and the technical requirements independent of a specific technological approach.

- Chapter 6 shows one implementation of the rights expression exchange framework that is naturally compliant with the general design described in Chapter 5. The components are capable of generating, wrapping, unwrapping, and interpreting rights expressions of the XML–based open digital rights language (ODRL)[4] [Ian02b] and are open for the support of any other current or future rights expression language. The implementation is coded in XOTcl[5] and reuses various software tools, such as the tDOM parser[6] for the handling of XML documents in general or the MySQL[7] data base server. The chapter gives detailed information about the tools class hierarchies and interfaces and presents examples for the tools' usage.

- Chapter 7 exemplifies an application of the rights expression exchange framework components. It describes the successful integration of the generator – and wrapper component in ActiWeb, a class library that is supporting extended web server functionality, and the deployment of the unwrapper – and interpreter component with an access control service.

- The conclusion gives a summarisation of this work. It highlights the developed findings in the field of exchanging and processing rights expressions. Additionally, the chapter addresses future work in this field that can be based on the work at hand.

---

[4]See http://www.odrl.net/
[5]See: http://www.xotcl.org/
[6]See: http://www.tdom.org/
[7]See: http://www.mysql.org/

Chapter 2

Digital Rights Management Systems

The electronic commerce of digital goods has significantly evolved in the last decade. When electronic goods are traded between geographically distributed consumers and providers, electronic contracts are concluded. An electronic contract is an agreement of two or more parties, on the exchange of rights to goods or services under certain terms and conditions. Electronic contracts differ from traditional paper contracts in their medium, i.e. in contrast to paper contracts, electronic contracts are/or can be digitised and exchanged via an electronic network.

A DRM system is facilitating the 'digital management of rights' [Ian01]. As stated above, electronic contracts comprise rights to goods and services and consequently are related to DRM systems. To draw the bridge to the economic aspects of DRM systems, Section 2.1 discusses the characteristics of digital goods and their business models. Subsequently, Section 2.2 gives some definitions of DRM and discusses their different perspectives. Section 2.3 introduces typical functions of DRM systems and describes a sample DRM system with its internal workflow. In Section 2.4 the chapter finally envisions a potential role of rights expression languages in DRM systems as a part of the next generation of e–commerce systems.

2.1 Trading Digital Goods

In this section the characteristics of digital goods and their business models are addressed. The section describes how digital goods are different from physical goods, and why this allows for new business models for digital goods.

2.1.1 Characteristics of Digital Goods

Digital goods are sometimes called electronic goods, information goods, virtual goods or intangible goods, as well as digital content or digital products. Digital goods sometimes also comprise digital services, such as digital phone or television services. In this thesis, the term digital goods will be used denoting both digital goods and digital services.

Digital goods are usually cheap in reproduction and cheap in distribution [SV99], as the customer pays for the storage and carriage. Advantages and disadvantages result for both consumer and provider: a cheap copy of a digital product keeps the reproduction costs low, and the cheap distribution via the Internet is also advantageous for the product vender. Nevertheless, these advantages count also for the customers who are able to easily copy and distribute digital goods (music files, films, etc.) to their friends. From the point of view of the music and video industry, this habit causes a decrease in sales of digital goods. However, to transfer digital goods via the Internet, they sometimes have to be compressed. Although also lossless compression mechanisms exist, the compression of digital goods may reduce the product's quality and consequently decrease the value for the consumer. Despite the sketched problems, economists consider the commercial potential of digital goods, which will be addressed in the subsequent section.

2.1.2 Business Models for Digital Goods

The digital medium offers a large spectrum of ways of commercialisation, and facilitates new, innovative business models for digital content [vWT03], i.e. the secure delivery of digital content will open global markets, reduce distribution costs, create more intimate contact with consumers, etc. [Rig03]

> "Licensing has become a familiar mechanism for providing access to some types of digital information (e.g. software), but is relatively new for other types (e.g. research journals). ... By

offering a distribution model different from that represented by copyright and sale, licensing has the potential to open new markets." [SD00]

Although the Internet is a driving factor for trading physical goods, this thesis is focused on technically supporting the formulation and processing of new business models for digital goods. In DRM literature the term *business model* is often related to different meanings. Osterwalder [OLP02] defines an e-business model ontology which is founded on four main pillars:

1. *Products and Services.* The products and services a firm offers, which represent a substantial value to the customers, and for which they are willing to pay.

2. *Infrastructure and Network of Partners.* The infrastructure and the network of partners that are necessary in order to create value and to maintain a good customer relationship.

3. *Relationship Capital.* The relationship capital the firm creates and maintains with the customers, in order to satisfy them and to generate sustainable revenues.

4. *Financial Aspects.* The financial aspects, which are transversal and can be found throughout the three former components, such as cost and revenue structures.

Following Osterwalder's definition for a business model, the subsequent paragraphs address electronic products and services and their financial aspects. In this thesis the term *pricing model* is used to express price and terms and conditions for a certain usage or access right to a digital goods produced by the seller, for example, 'listening 50 times to the music file X for the payment of €2.00'. Issues concerning the importance of the *infrastructure and network of partners* is generally addressed in Section 1.2.

Ten years ago, for buying song 'S' a customer had to purchase the entire album for about €10.00 in form of a compact disc or on vinyl. The pricing of that time allowed only one purchase condition for song 'S', because it was bound to the storage medium of a compact disc. Producing a CD for each song on the album would have been far too cost intensive and risky. Gainful business models for digital goods vary depending on their product type (or *dimension* [Kop99]). For example, usually, digital information containing the weather forecast or the last weekend's baseball scores are valuable for the customer only once and only at the right time. A weather forecast for January 1st 2004 will be of no value for the customer on January 20st 2004. In contrast to this, the desire to listen to the same music file over and over again can be very valuable for the customer. Therefore, the conditions for reading the latest news will be different from the conditions for listening to music files. For example, a reasonable pricing model for an online newspaper, whose value is determined by timeliness, is a monthly subscription for a fixed higher price or a lower price payable for each access. A sensible pricing models for a music file might be unlimited access rights to that music file for a higher price or a lower price for access rights to that music file limited to five times. Generally, a sensible business model is certainly to give away the digital goods for free without any restrictions, and finance this business by other income streams, such as advertisement [SV99]. But as such business models lack contracts on digital goods and also exchanging and processing rights expressions, these business models are not further investigated.

To understand the potential of digital goods and to find the right business models, different dimensions (or product types) for digital goods have been identified by Choi et al. [CSW97] and Koppius [Kop99]. Choi et al. name the five dimensions *transfer mode, timeliness, intensity in use, operational usage,* and *externalities*. Koppius has further developed the work of Choi et al. and additionally distinguishes consumer (buyer) dimensions, such as *value determination, perishability, recipient, complexity of product use, externalities* and provider (seller) dimensions, such as *specifiability, customisability, substitutability, intensity in use,* and *existence of a tangible equivalent*. Furthermore Koppius defines the delivery process–related dimensions *transfer mode* and *options for tangible support*. The forming of the named dimensions influences the 'right' business models for digital goods.

As soon as customers decide to purchase and electronically consume digital goods, there is a need to express participating parties (consumer and seller), price, terms and conditions in electronic contracts. Electronic contracts can be formulated in *rights expression languages* (RELs) that will be described in more detail in Chapter 3.

2.2 Digital Rights Management (DRM)

This section gives a short introduction to the term *digital rights management*. It provides and discusses an extract of DRM definitions, concepts and visions which will be used in later sections. The section also introduces the different perspectives of DRM, such as the legal, functional, and technical perspective and points out the perspectives that are addressed with this thesis.

2.2.1 DRM Definition

Today, the term *digital rights management* is frequently associated with online music shops, secure viewers, and copy protection mechanisms. But these areas only are a small part, respectively a particular type of DRM system applications. Digital rights management has been defined by various researchers and industry associates. To give an overview of the DRM spectrum, here are some recent definitions:

> Iannella: ".. Digital Rights Management covers the description, identification, trading, protection, monitoring and tracking of all forms of rights usages over both tangible and intangible assets including management of rights holder's relationships." [Ian01]

> Gunter et al.: ".. DRM systems enable sellers of digital content to move beyond current distribution." [GWW01]

> Neylon: "The technical means by which content is dynamically licensed for, or protected against, a particular use is known as *digital rights management*. The tools of digital rights management do not define how commerce must be conducted. Rather, they allow business models to be defined and support their implementations." [Ney01]

> ContentGuard, Inc.: "The reality of the Internet and the need to control the use of digital content and digital services has fueled the development of technologies that attempt to manage, secure, control, and automate the flow of content and the access of services. Digital Rights Management (DRM) is the common term associated with such technologies. Other technologies such as Digital Asset Management, Content Management, and Trust Systems are also getting incorporated into the DRM workflow.

The DRM space is becoming more important and, in many cases, required to enable certain business models." [Con00]

The various definitions show the broad spectrum of DRM. The development of DRM has emerged from the commercialisation and trading of goods and services over the electronic/digital medium. The DRM definition that is used in this thesis is introduced in the subsequent chapter. Iannella states that: *"it is important to note that DRM is the "digital management of rights" and not the "management of digital rights". That is, DRM manages all rights, not only the rights applicable to permissions over digital content* [Ian01]. However, the definitions also indicate that DRM systems can be seen and influenced from different perspectives, as the subsequent section shows.

2.2.2 Perspectives of DRM

The four definitions mentioned above of DRM differ from each other not least because the authors are looking at DRM from different perspectives. The definition by Iannella lists the functions that are required for a sophisticating and successful DRM system. The definition of Gunter et al. does not specify the functionality in detail, but sees DRM as a means to extend current distribution channels – it describes the intentional perspective of DRM. That means, to comprehensively describe DRM systems several perspectives need to be considered. After reviewing a large number of DRM definitions, I have identified six different perspectives that give an overall view on DRM systems: the *intentional perspective*, the *technical perspective*, the *functional perspective*, the *legal perspective*, the *social perspective* and the *economic perspective* (Figure 2.1).

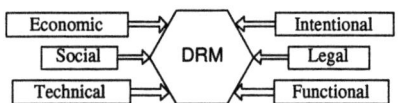

Figure 2.1: The six perspectives of DRM

The six perspectives of DRM influence each other. A change of the *intentional perspective* of DRM has an impact on the functions and also on the technical implementation of the DRM system. This example shows that some perspectives dominate others. A change in the legal, social, intentional, and economic perspective causes changes in the functional and technical perspectives. The mutual relationships between the different perspectives can be very complex and are subject to future research. In the following sections each of the identified DRM perspectives is described in more detail.

- *The intentional perspective.* The intentional perspective defines the overall goal for a particular DRM implementation. One intention of a DRM system might be to protect the property rights of an enterprise's asset. The asset in this case could be tangible (e.g. a book) or intangible (e.g. an Ebook). Another intention of a DRM system might be to establish the awareness of intellectual property rights (IPR).

- *The economic perspective.* The economic perspective shows how economic factors, e.g. the business model or the market environment, can influence a DRM system. Neylon states that "the tools of digital rights management do not define how commerce must be conducted. Rather, they allow business models to be defined and support their implementations" [Ney01]. Accordingly, the implementation of DRM systems depends on the underlying business models, i.e. the economic goal of the system. For example, the definition and enforcement of usage rights are less important for a platform like 'The Knowledge Conservancy'[1] that provides any material "free-to-read", than for a commercially run electronic market for music tracks like MusicNet[2] and Apple's ITunes platform[3].

- *The legal perspective.* Amongst other potential applications, a digital rights management system can provide the technical environment to protect intellectual property rights. Intellectual property rights are defined by the law. So law influences digital rights management system with respect to compliance, investigation, and enforcement mechanisms. The legal perspective represents the influence of these predominant laws on a DRM system. For example, copyright law could be adjusted in the near future to be digital age compatible. These changes would most presumably require a reengineering of the existent DRM systems. Copyright law in the USA and Europe differ from each other in their national interpretation or even collide. For a globally operating DRM system the legal perspective will play an important role since it is a considerable challenge to integrate different

---

[1] See: http://yen.ecom.cmu.edu/kc/
[2] See: http://www.musicnet.com/
[3] See: http://www.itunes.com/

legislations of various countries. The *Creative Commons*[4] is an initiative that aims at formulating a set of copyright licenses according to copyright law in order to help content creators to distribute their protected works. A DRM system that wants to make business with those content creators will probably have to support those licenses.

- *The social perspective.* Crucial success factors of today's electronic markets or Internet platforms are their acceptance and frequency of usage. These factors are governed to a great deal by social norms. The *social perspective* addresses social, personal, and psychological aspects of a DRM system. Clients must have incentives to be willing to use platforms with a digital rights management. The social perspective addresses questions like: "Why should a client use this platform with DRM?" – because it is convenient or a good information source? Is it maybe because the client is aware of intellectual property rights (IPR) protection? Why should the client acquire digital content of the platform with DRM, where content is liable for costs or at least requires registration instead of searching for a free, anonymous source? DRM systems are influenced by the public sense of IPR protection. Consumers who understand the risks associated with pirated electronic content will more likely acquire content from legitimate sources with digital rights management [oAPI01].

- *The functional perspective.* The functional perspective describes the functions of a DRM system, e.g. protection, management and monitoring of property rights, enforcement of terms and conditions, creation and management of contracts, revenue stream control et cetera. The technical perspective is vastly influenced by the other perspectives of the DRM system. The functionality of an exchange platform will be designed for example depending on the platform's intention or business model. If the business model of a platform changes, the functions of the platform will change as well.

- *The technical perspective.* The technical perspective has a number of sub-areas. It covers, for example, the data model, the secure electronic environment, the system architecture, the applied standards, the protocol stack, the authentication and identification mechanisms, and the digital rights language. The technical perspective is also strongly dependent on all other perspectives. Changes in all other perspectives require most likely technical reengineering.

---

[4] http://www.creativecommons.org/

Figure 2.2 illustrates all identified DRM perspectives in the order of their influence on the technical implementation. The characteristics of each perspective finally determine the *strategy* of an actual operating DRM system. A perspective on the higher level has influence on the perspectives on lower levels. Consequently, if the economic perspective of a DRM system changes, the functional and technical perspectives change as well. For example, if in the economic perspective it is specified that only rights to physical goods shall be managed, no functionality that supports access control to digital resources is required. If a change in the higher perspectives is made, this change will cause costly modifications in the lower level, i.e. reengineering of the functional, respectively technical implementation. This thesis focuses on the lower namely the functional and technical perspectives of DRM systems. It aims at providing technology to facilitate economic functions or concepts (e.g. to express various pricing models) and supporting the requirements of the legal perspective (e.g. providing digital signature information), as well as giving new functionality to DRM systems, such as interoperability.

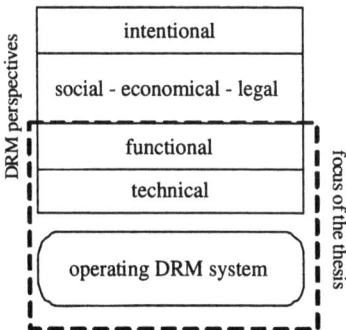

Figure 2.2: The DRM perspectives in the order of their influence on DRM systems

The DRM perspectives introduced in this section help to classify and describe DRM systems. To provide an overall specification of a present or future DRM system, all named perspectives should be considered. The definitions given in the previous section only partly address DRM systems. For example, Iannella and Neylon mention the functional perspective of a 'typical' DRM system, Gunter names one particular economic goal that can be achieved by using DRM systems, whereas ContentGuard addresses both the functional perspective by naming typical functions, such as "manage, secure, control and automate the flow of content", and the economic perspective by "DRM ... is ... required to enable certain business models."

In this work the following definition for DRM will be used:

> "DRM systems provide techniques and functions that facilitate the digital management of rights to digital or physical goods or services. Each actual DRM system thereby may aim at addressing different economic, or social goals, respectively implement legal requirements."

2.3 A Sample Digital Rights Management System and its Functions

This thesis deals with the exchange and the processing of rights expressions. Rights expressions are exchanged between system components of DRM systems or between DRM systems. Therefore, in this section a sample DRM system along with its basic functions, system components, and internal information flow is introduced. To describe a sample DRM system and typical DRM functions various prevailing DRM systems have been analysed. For the analysis the DRM systems described in [RTM02, IBM02, Dig03, Mic03, FFSS01, KM00, Nok01, DK01] have been consulted. Accordingly, the sample DRM system described in this section is a non-existing system which includes the typical characteristics of the analysed systems and may serve as a reference DRM system.

The analysis shows that there are many variants of DRM system architectures, which makes it difficult to describe a 'typical' DRM system. However, DRM systems can be characterised by their basic functionality. Therefore, this chapter starts with a brief look at DRM systems from the functional perspective and identifies their basic functions, namely *Content Provision, Content Safekeeping, Offer Placement, Content Preparation, Content Distribution, Booking, Payment, Authorisation, and Content Consumption*. The functions identified are provided by DRM system parties (or components) such as the content provider, DRM platform, etc. All analysed DRM systems can be described on the basis of these functions, although all these systems have different technical architectures. The sample DRM system described in Section 2.3.2 covers all identified basic functions and displays a typical architecture in which customer, DRM platform, content provider, and clearing house interact. In Section 2.3.3 the information flow through the sample DRM system is described. The chapter closes with an introduction of some commercial DRM systems with respect to their functions and system components; it finally addresses DRM system designs which differ architecturally from the sample DRM system. In some of these systems, additional parties come into play and assume responsibility for one or more DRM functions, thus changing the DRM's architecture and information flow.

2.3.1 DRM System Functions

A DRM system provides the digital management of rights to digital or physical goods or services. The 'digital management' involves several functions, such as the provision of goods or services, their distribution, purchasing, and the delivery or rendering respectively consumption. The occurrence of such functions describes the functional perspective of DRM systems. All DRM functions require the deployment of suitable security mechanisms. The provided functions of a DRM system are more or less similar, but the ways they are implemented vary, which means that similar DRM functions are executed by different system components with varying responsibilities and differing system architectures.

For example, let us assume a customer wants to access secure digital content. A license (see Section 4.3.1) specified by the content provider defines the rights governing access to the content. Both the content and the license have to be delivered to the customer, the rights have to be interpreted and executed, and the content has to be rendered. The implementation of these functions can differ with respect to the following questions: Are the license and the secured content delivered to the customer together or separately? Are access rights interpreted and enforced by a mobile software agent, by a secure viewer on the client's PC, or possibly by a web server which regulates access to its realms? The mentioned variations are different technical implementations of the same functions. That is, these different implementations or technical architectures describe the technical perspective of DRM systems.

In this subsection, basic and extended DRM functions that have been identified in the analysis are introduced. Identifying DRM functions helps us to categorise and describe DRM systems. The functions introduced are

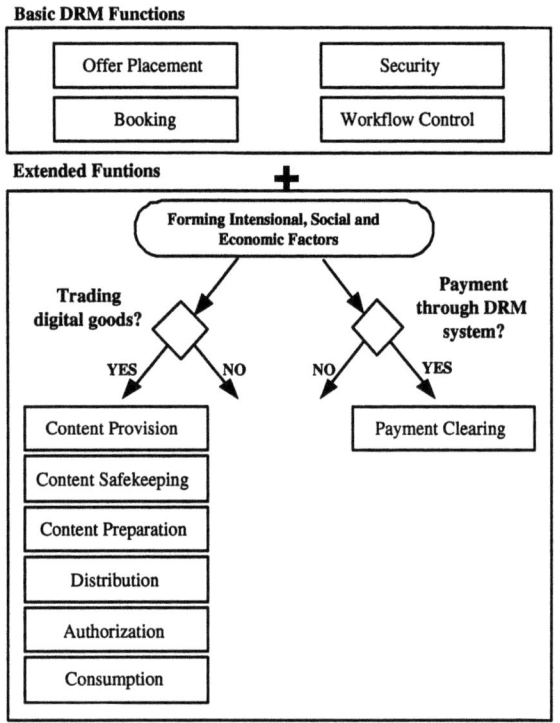

Figure 2.3: Basic and extended functions of DRM systems

used in the subsequent sections to describe the sample DRM system as well as current DRM system implementations. Basic functions are *Offer Creation, Booking, Security* and *Workflow Control*. Whether a DRM system provides extended functions, depends on the forming of the system's intentional, social, and economic perspective (see Figure 2.3), i.e. the DRM system's strategy. For example, if the DRM system trades digital goods the functions *Content Provision, Content Safekeeping, Content Preparation, Distribution, Authorisation,* and *Consumption* should be supported as well. If monetary transactions are part of the exchange process, DRM systems have to provide a *Payment* respectively *Clearing* function or at least provide an interface to such components. Examples for DRM systems where rights are exchanged but payment is not an obligatory part of the transaction are educational projects such as Universal[5] or COLIS[6], both of which are brokerage platforms for learning resources. The list of extended DRM will probably get longer as new generations of DRM systems emerge.

Each of the following DRM functions comprises a number of *DRM activities*. The activities differ depending on the technical implementation of the respective function. For example, the consumption of digital goods may include the following activities: handling the access request of the consumer, authenticating and identifying the consumer, identifying the resource that is subject to the access request, decrypting and decompressing the digital good(s), granting or denying access to the digital goods according to the license, and rendering the digital goods according to the granted permissions in the license.

- *Content Provision*: Content providers, e.g. sellers of digital music, videos or Ebooks [Ass00, oAPI01], who decide to distribute electronic goods via a DRM system have to make the content available to the DRM system, that is, e.g. encoding the digital content in the right format (e.g. AAC+ in MP4) and securely uploading the digital content to the DRM system. This initial function is called *content provision*. Content provision also includes the delivery of content metadata, such as workflow metadata[7], metadata on security (e.g. using XML–Signature [BBF+02]), and product metadata (e.g. using Dublin Core [Dub01], or learning object metadata (LOM) [IEE02]) for content discovery. Metadata provision can be classified as an extra function. Some DRM systems do not consider content provision to be a basic function and assume that content is simply available on the DRM platform.

- *Content Safekeeping*: Content safekeeping (or administration) deals with making the content available to the DRM system. This function merely supports the secure storage of traded content.

- *Offer Placement*: An offer contains the terms and conditions also called usage and access rights, which regulate content usage. The offer placement function provides a means for the content provider to specify these terms and conditions.

---

[5]See: http://www.educanext.org/
[6]See: http://www.colis.mq.edu.au/
[7]See: The Workflow Metadata Initiative, http://www.metadata.sis.se/

- *Content Preparation*: In content preparation, the content is transformed into a secure, tradable format. The result of this process is a format called a *secure container*. The form of this container varies in the different DRM systems. A variety of security technologies are used to create containers, and their ingredients vary from system to system. For example, in some systems the containers comprise the digital content and its access rights, in other systems the access rights are transported separately from the content. Content preparation also included the technical bundling of digital content.

- *Booking*: The booking function provides services for the customer to purchase content or, more precisely, to purchase usage rights for content. Booking or purchasing the digital product asks for a contract between the content provider and the consumer. The contract should have an exchangeable and standardised format, and ideally it should be written in a rights expression language (see Chapter 3). Booking also includes reporting for the content providers that want to know how often their content was requested.

- *Content Distribution*: The content distribution function provides secure distribution channels to the customer.

- *Payment/Clearing*: In a great number of contracts, the purchase of digital content requires a payment from the consumer to the content provider or to the DRM platform (acting as a proxy). The payment has to be executed according to the specifications in the contract. For this task, a clearing house is required which provides various payment methods (credit card, debiting, electronic cash, etc.), maintains accounts for all involved parties, and facilitates the settlement of payments. To learn more about electronic payment systems, please refer to [SS03].

- *Authorisation*: Authorisation to access or use goods or services is allowed by a *token*. It is important to note that the token is not the specified license. In this thesis, a token is defined as *a technical means*, such as a decryption key for the secure container, which enables the customer to use the content according to the license. Depending on the security strategy, the token is either transmitted from the platform to the customer or is hosted by the secure viewer.

- *Content Consumption*: Content consumption provides mechanisms to access and render the content kept in the secure container. Often, consumption is facilitated by a DRM client software on the consumer's computer [vWT03, Mic03, IBM02], called secure viewer. The content consumption function also facilitates *content tracking*, which might be a relevant activity for IPR protection.

- *General Function: Workflow Control*: As various components interact in a DRM system, each component requires the integration of a workflow mechanism to control and coordinate the sequence of tasks and activities in the workflow through a DRM system.

- *General Function: Security*: The DRM system processes digital content and data that has to be constantly protected. The content and data (e.g. contracts) have to be protected against various types of fraud, such as unauthorised access or the modification of rights information in contracts. The following security techniques are used in DRM systems.

  - *Encryption/Decryption*: Most DRM systems use encryption and decryption to protect the data circulating in the system. For efficiency reasons, symmetric key algorithms are generally used to encrypt and decrypt the digital content, while asymmetric key algorithms are used to generate digital signatures, establish secure channels, and to encrypt and decrypt symmetric keys. Thus, encryption and decryption support the DRM functions *Content Distribution, Content Preparation, Content Consumption*, etc. and facilitate digital signatures and secure containers.

  - *Digital signature*: Digital signatures provide a means of verification, integrity checking, authentication, and non-repudiation. For example, digital signatures can be used in the *Offer Placement and Booking* function to evidence their validity.

  - *Watermarking*: Watermarks bind information directly to the content. Most watermarking technologies claim to be unremovable from the content (even after data compression), which enables the lasting identification of digital content. Thus, watermarks support, for example, the *Consumption* function, when an access request to a certain digital content has to be checked against a licence. For more detailed information on watermarking in DRM systems, refer to [Pet03].

  - *Secure Container*: The secure container technique is used as a secure transport format for the distribution of digital content. The container protects the content from unauthorised access. Erickson [Eri01] states that the role of secure containers (wrappers) is that of a mediator service. The wrapper can link to services such as the repository, authentication and authorisation.

- *Public Key Infrastructure (PKI)*: PKI is the basic infrastructure for many security technologies [KL89]. It is used to facilitate digital signatures, encryption and decryption services, secure transport channels, key registration, certificate issuing, revocation services, etc. Thus, PKI supports various DRM functions, such as *Authorisation, Offer Placement, Content Consumption, Booking*, etc.

- *Proprietary Mechanisms*: Not all DRM systems use standard technologies to ensure system security. Some systems use proprietary mechanisms and processes, for reasons such as unsophisticated standards in a particular field or the fear that known technologies are easier to circumvent.

The design of the security concept can strongly influence the entire system architecture and information flow.

Functions versus System Components. Previous studies in this field deal with the definition of DRM system components rather than functions; for example, Rosenblatt, Trippe and Mooney [RTM02] define a DRM reference architecture on the basis of standard components. In my view, it is more transparent to understand, evaluate, compare, and categorise DRM systems using a set of functions. Whereas functions describe the smallest unit of the DRM system (module), a system component comprises several functions. The functions introduced can be combined in many different variants. Some might even be processed by hardware components; for example, the European pay TV contractor Premiere World[8] uses smart cards to handle parts of its security process. In the next section, one of many possible combinations is introduced.

DRM system implementations that support the identified functions are sometimes called *DRM middleware* [FFSS01]. Many businesses may deploy DRM middleware, with possibly different products, information base and privacy policies. This makes DRM systems different from specific e-commerce solutions, such as PressPlay[9], which uses the Microsoft DRM

---
[8]See: http://www.premiere.de/
[9]See: http://www.pressplay.com/

system, or MusicNet[10] which operates with the Real Networks' DRM system (see also Section 3.5).

2.3.2 A Sample DRM System

This section introduces a sample DRM system based on the functions referred to in the previous chapter. As mentioned above, the sample DRM system is a non-existing system, but comprises the typical characteristics of the analysed systems and may serve as a reference model for DRM systems that manage payments and deal with digital goods, such as music and video files, Ebooks [KF02], images, etc.

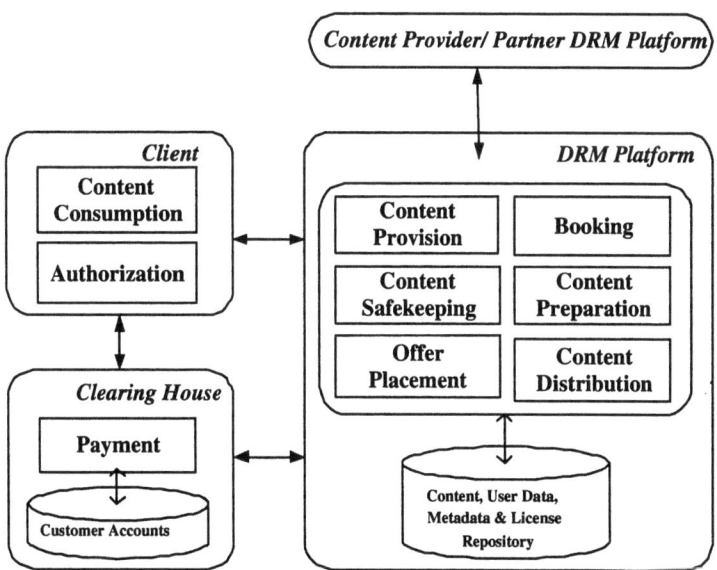

Figure 2.4: A sample DRM system

Figure 2.4 shows a DRM system that comprises three components: the client (or consumer) who desires to consume digital goods, the DRM platform, and the clearing house. The three components are interoperating with each other. The content providers and partner DRM platforms represent parties that interact with the DRM system but do not provide DRM

---

[10]See: http://www.musicnet.com/

functionality. The DRM platform is the key component which controls the information flow through the sample DRM system (see Section 2.3.3) The information flow integrates all participating components of the DRM system (content providers, consumers and the clearing house). The DRM platform provides the functions content provision, offer placement, content safekeeping, content preparation, content distribution, and booking (see Figure 2.4). The payment function has been outsourced to a clearing house. Content consumption and authorisation are supported by DRM client software on the consumer's personal computer. The subsequent chapter describes the process of a typical DRM process and shows the information flow through the sample DRM system of Figure 2.4.

2.3.3 A Sample DRM Process

This section exemplifies a DRM process that undergoes all functions in the DRM system – from content provision to content consumption. In the analysed systems, which all support the management of rights for digital goods or services, the DRM process is implemented either the same way or alike as shown below.

1. *Content Provision.* First of all, content has to be provided by the rights holder (see Figure 2.5). Content provision can be technically implemented in many ways, for example by uploading to a content server or by sharing a folder on the provider's computer. An interface has to be provided for manual provision by content providers as well as for automatic provision by cooperating DRM systems. In order to facilitate interaction between cooperating DRM systems, a standardised interface is necessary. During the provision process, the content has to be protected from unauthorised access by security mechanisms, for example by a secure channel (e.g. with the help of SSL), or by encrypting the content. The content metadata can be provided separately from the content. A graphical user interface should be provided for the manual input of workflow, security, and resource metadata by content providers, and/or a standardised format could be offered for the automatic provision of content metadata records.

2. *Content safekeeping.* Once the content has been provided, it is stored in a secure environment in the content repository. Depending on the DRM system concept, the content is stored in plain format, or in a security wrapper (secure container). The metadata is stored in the metadata repository.

3. *Offer Placement.* Content providers offer their content on certain terms and conditions. In this sample system, these conditions are not fixed but can be defined individually for each unit of tradable content. Specifying these terms and conditions can also be regarded as rights metadata provision. The provision of rights metadata results in an offer. The offer placement function has to be flexible and various business models shall be supported. In practice, the content providers are guided through a menu where they are able to specify terms and conditions for any of their resources. Similar to product metadata, the offer can either be provided personally by the content provider or by a cooperating DRM system acting on behalf of the content provider. In the latter case, an interface has to be provided to receive and exchange offers and to process them automatically. The licenses are stored in the license repository.

4. *Booking.* When consumers wish to purchase content, they will need to contact the DRM platform. The buying desire usually precedes promotion activity. The connected e–commerce system is in charge of such activity. The booking module merely should support searching for content and browsing for offers. However, the DRM platform's booking module then receives the customer's purchase request and returns terms and conditions, as well as information on the payment process to the customer. The customer agrees to the terms and conditions of the platform by signing the offer previously defined, or rejects it by not doing so. Accepting the offer results in a legally binding digital contract. The contract is formulated using a machine readable rights expression language (see Chapter 3).

5. *Payment.* The customer then contacts the clearing house and initiates the payment process. The clearing house balances the customer's and the platform's accounts and notifies the DRM platform of the payment. The electronic payment system *PayPal*[11], which is currently used predominantly by online auction participants, supports this payment procedure. However, other payment systems could be used as well.

6. *Content Preparation.* As soon as the booking module receives the payment notification, the content has to be prepared for distribution. In this sample system, this includes the following steps:

---

[11]See: http://www.paypal.com/

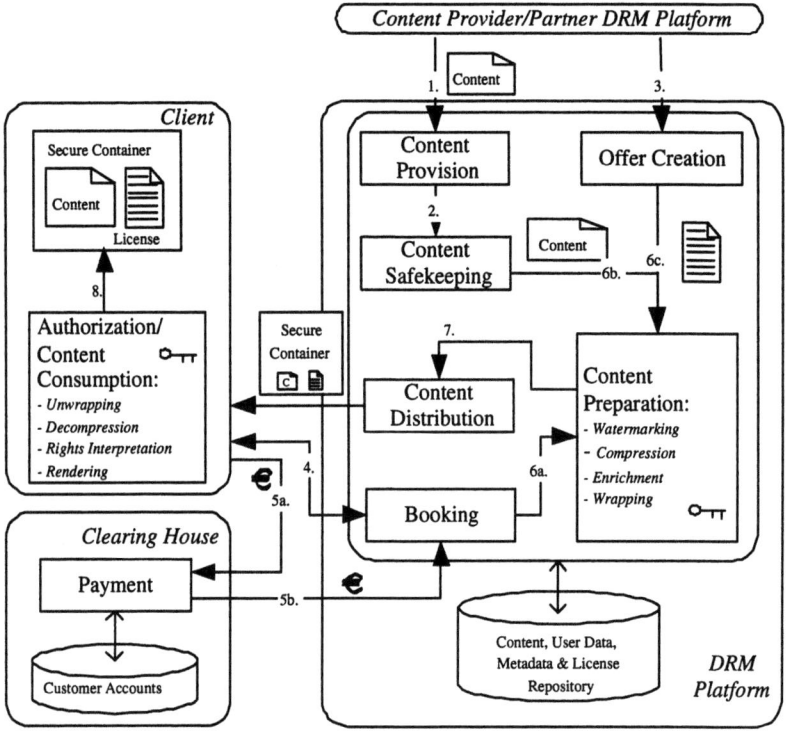

Figure 2.5: A sample DRM process

- *Watermarking.* A watermark is added to the content. This watermark comprises metadata on the content and enables, e.g. the identification of content.
- *Compression.* The digital content is compressed into a manageable size, e.g. from original memory–intensive picture representation to JPEG format.
- *Metadata Enrichment.* The digital content is now enriched with metadata, such as licensing, product, security and workflow information. This metadata is restored from the platform's repository. This system is designed send the license information directly with the content (more about the difference between electronic contract and license in Section 4.3.1). The licensing information

is retrieved from the concluded contract between consumer and platform. The licensing information is available in a rights expression language, and comprises concrete usage and access rights of the consumer to the digital content; for example, the content may be accessed by one particular consumer in the read–only mode before 31st December 2004. The license is digitally signed by the platform.

- *Wrapping.* In order to protect the content against unauthorised access the meta data (including the license) and the compressed content are wrapped by means of a symmetric key mechanism. The result of content wrapping is the secure container, which constantly prevents unauthorised access to the content.

7. *Content Distribution.* The content is now delivered by the platform to the customer for consumption. In the sample system the content may be distributed via an unsecured network, as the content is protected by the 'secure container' technology. The distribution channels should be able to serve various kinds of end devices, such as PCs, PDAs, cellular phones, etc.

8. *Authorisation/Content Consumption.* In this system design authorisation and content consumption are executed by the DRM client software that has to be installed on the customer's PC. The DRM client is trusted by the DRM platform and naturally owns the authorisation token to access the secure container and thus is able to render the content in accordance with the the terms and conditions in the license. The authorisation token, in this case is the symmetric key with which the secure container can be unwrapped. In this process it is assumed that the symmetric key has been exchanged before hand. Alternatively, the decryption key can be sent separately (separate delivery). The DRM client is sometimes referred to as a *secure viewer*. The consumption process of a customer's access request includes the following steps:

- The secure container content is *unwrapped.*
- The compressed content has to be *decompressed.*
- Now the secure viewer has access to the content and the license, among other metadata. The license, formulated in a rights expression language (REL), is parsed and *interpreted* by a REL interpreter that is part of the secure viewer. The secure viewer

queries the license by the application programming interface (API) of the REL interpreter for the data in the license. If the license allows the requested access, the secure viewer releases the content for rendering. Advanced functions, such as *quality control mechanisms* for ensuring content quality after encryption, compression, transmission and decryption of the digital goods, could be applied at this time. Prior to rendering the content, the client software also has to execute a number of *security checks*. For example, it checks the digital signature of the license to see whether the license has been manipulated during the distribution phase. The secure viewer verifies that the content identification number in the license is identical with that in the watermark, etc.

- Finally, the client *renders* the content in compliance with the license specifications (a process which is also called *rights enforcement* [GSZ03]).

The consumption of usage rights to a web site is handled in a different mannor from the process described above. In that case, access rights to a web site are enforced e.g. by a web server that protects the web site from unauthorised access rather than by a secure viewer on the consumer's computer. Also if the digital goods to be protected are web sites, they usually do not have to be decompressed and encrypted prior to distribution respectively delivery. An example of such a DRM system is the German online newspaper 'Spiegel Online'[12], where dossiers on certain subjects can be purchased and then accessed via the web browser. In addition, DRM systems are not only prevalent in business–to–consumer relationships, but also in business–to–business relationships, where DRM is used to regulate trading among electronic brokerage platforms [GSZ03]. This means that the consumption or the trading of digital goods may have various facets or technical implementations. DRM systems must provide consumption mechanisms for all formats in which content is offered. The various technical mechanisms that are provided by a DRM system form the DRM system's technical perspective.

Digital content is not always directly sent to the customers. In some applications, such as in Nokia's distribution of ring tones [Nok01], the content is delivered by superdistribution. Superdistribution is an alternative distribution channel where digital goods can be given away freely without resistance from either copy protection or piracy. The originator never relinquishes ownership rights with the digital good when distributed [Cox94].

---

[12]See: http://www.spiegel.de/

Superdistribution can be implemented for example with peer–to–peer networks or by people who exchange resources among each other, i.e. potential customers exchange the secure containers privately in unstructured ways. These containers do not contain a license. If the customer decides to purchase a ring tone, s/he requests the license and token at the DRM platform belatedly. A sample DRM system that describes the superdistribution process can be found in [Gut03].

In terms of security, the process described above only sketches a few mechanisms used in the field of DRM. The usage of other security mechanisms changes the information flow through the DRM system. Every DRM system requires a system for the global or at least system–wide identification of digital goods and users. Thus, the usage and access rights of certain users can be uniquely assigned to certain resources in rights expressions respectively licenses and contracts. This unique identification then allows the correct fulfillment or processing of licenses respectively contracts. Such identification mechanisms are, e.g. DOI [Nat00] for the identification of digital content and x509 certificates [IT93b] for the identification of individuals.

The digital goods or services traded via the DRM system are related to metadata that further describes the resource. The DRM system requires the use of a standard description language, such as dublin core [Dub01] and/or LOM [IEE02], that eases the recording respectively exchange of such data. Finally, an infrastructure for security services, such as PKI [KL89] has to be provided. The process described above does not address the technical protocols that the DRM components use for communication. To read more about this technical detail, please refer to Erickson [Eri02] who describes a reference architecture for the communication between DRM system components on the protocol level. Erickson introduces the required protocols and standards in context of the 'DRM reference architecture' described in [RTM02]; this DRM system architecture has already been addressed in Section 2.3.1.

2.3.4 Commercial DRM Products and DRM System Variants

The sample system in the previous section describes one possible form of a DRM system as well as one typical DRM process through such a system. In this section, actual implementations of DRM systems are described with the help of the functions identified in Section 2.3.1 to show that those hold for all investigated DRM systems. In the DRM process of Section 2.3.3, the license that grants access is packaged with the resource in a secure container (combined delivery). In other DRM systems the resource comes in the secure container but the license and the token have to be acquired separately from the DRM system (separate delivery). Both approaches have advantages and disadvantages. The license which is bound directly to the content reduces the complexity of security and communication in the DRM system because only one secured transmission between consumer and DRM platform is necessary. The drawback of this approach is that the license cannot be changed once it is issued and integrated into the secure container, e.g. in cases where access respectively usage conditions change over time and outdated versions of the secure container are still circulating. If the license is distributed separately from the content, an additional tamper–resistant connection to the DRM system is required in order to receive the license, but this approach makes the DRM platform very flexible in controlling, varying, and changing the terms and conditions for the digital content and allows superdistribution. Every time use or access rights are purchased for the packaged product, the consumer has to contact the DRM system that in turn sells a license under the current (changed) conditions. However, each of the two mechanisms have sensible applications; therefore a DRM system should support both.

InterTrust has done pioneer work in the field of DRM. In the description of InterTrust's DRM system below, the terminology and the graphic symbols from the previous sections will be used. In the InterTrust system design [DK01], the license and the content are handled separately from each other, both in a protected format. The licenses are administered by an additional, independent component called the Content Rights Server (see Figure 2.6). The booking and payment functions aer delegated to an external e–commerce system. Once the customer has settled the payment with the e–commerce system, the authorisation module (called the Authorisation Generator) sends an authorisation (token) to the customer, who can then use it to retrieve a license for the purchased content from the Content Rights Server. Content consumption is then processed by the Rights|System Client. InterTrust, does not provide a DRM middleware implementation. However, InterTrust has recently had success in licensing its DRM specifications.

The Windows Media Rights Manager [Mic03] differs from the sample system introduced in this chapter in that the DRM platform does not host the booking service. The clearing house is responsible for the booking process. An additional booking module which challenges booking requests has to

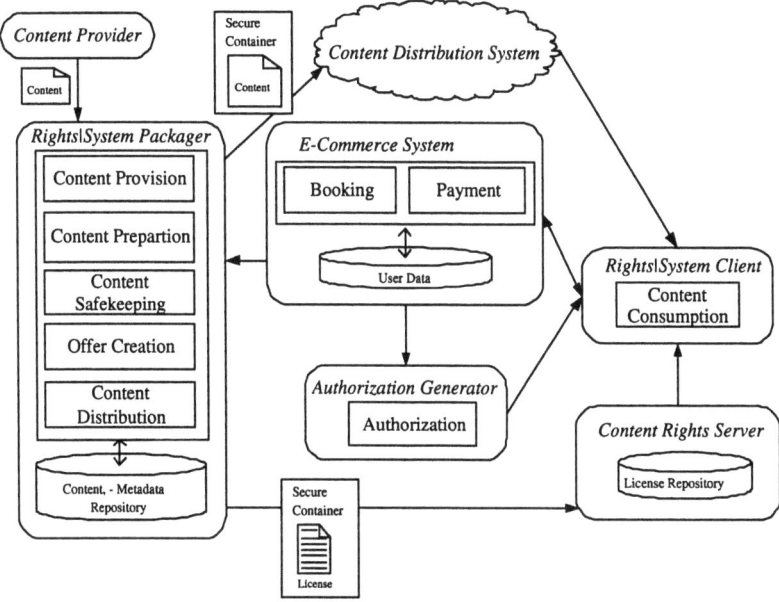

Figure 2.6: InterTrust's DRM system

be installed in the clearing house component (called Microsoft's License Server). As Microsoft's system delivers the licenses separately from the content, the booking module is also responsible for delivering the license to the customer once the payment has been made.

IBM's Electronic Media Management System (EMMS) [IBM02] distributes content and the associated rights together in a secure container. With the exception of its payment function, this system resembles the sample DRM system. However, EMMS can be integrated into e–commerce systems which provide distribution and clearing services.

ADo^{2}RA is a DRM system developed by Digital World Services (DWS) [Dig03], which is part of the Bertelsmann Group. The system is designed with an separate component for the majority of DRM functions. It is worth noting that ADo^{2}RA uses a sophisticated two–step solution to facilitate the DRM function "authorisation": The content as well as the usage and access rights that have been purchased by the consumer are not delivered directly to the consumer but stored in a *rights locker*. The rights locker is a central repository for content and tokens which is accessible to the customer from various locations, e.g. from the car, home, or office, and from different device types, such as mobile phones, PCs or PDAs. To access the rights lockers from their current location customers have to authorise themselves to the rights locker. This approach has the advantage that content and rights do not need to be ported from one device to another by the customer.

The following approaches have not yet been developed as commercial DRM systems but introduce alternative technical approaches to the implementation of DRM systems.

Feigenbaum et al. in [FFSS01] introduce a very generic, distributed system architecture in which separate components are responsible for each function, including that of content safekeeping, packaging, and authorisation. This approach is also designed with a rights locker for the storage of the customer's licenses. One prerequisite for the implementation of a rights locker is the separation of content and licenses.

The study of Konstantas and Morin [KM00] presents an agent–based approach in a DRM system developed as a prototype. In this approach, the content provider is responsible for provision and offer placement, and delivers the content as well as the respective license to an agent platform via a secure channel. The agent platform then takes care of the content preparation. The content and the license are wrapped together within an agent. This agent is the only application that is permitted to access the content; thus the agent permanently secures the content. Agents containing digital content can be released through the common distribution channels. They also provide content consumption functions. In order to be executed on the customer's PC, agents require a suitable agent platform, for which Java technology was used in this prototype. Prior to accessing the content, the consumer has to consult the clearing house, which is responsible for booking and clearing. The customer obtains access (in the form of a token) from the clearing house and consequently does not directly get in touch with the content provider. The clearing house transfers payment and booking information to the DRM platform on a regular basis.

2.4 The Role of Rights Expression Languages in DRM

Section 1.2 highlights the importance of collaboration and compatibility in the field of DRM and explains the need for standardised contracts. The technical means to provide such collaboration and compatibility is to work with standardised technology. Such technology is sometimes developed by the industry or can be adopted from standard–setting bodies, such as the World Wide Web Consortium (W3C)[13]. A prerequisite for collaboration is that all participants agree one a set of technologies.

The need for collaboration occurs throughout the entire supply chain (see Section 1.2), including collaboration among DRM systems (in the B2B and the B2C context) and among DRM system components. Generally, the focus of this thesis, is the exchange and processing of rights expressions between distributed systems, either DRM systems or DRM system components. Various DRM systems have different representations of rights expressions respectively contracts. For example, in one system the contract information is stored as database entry with a fixed number of fields (e.g. on the EducaNext platform[14]), in another system the contract information may be stored in software objects, and in a third system contracts and their content can be kept as rules and facts in an contract expert system. To exchange contracts or rights expressions with other DRM systems or DRM system components with customers or third parties, contracts or rights expressions have to be formulated in a standardised format. Such a format is offered by rights expression languages (see Chapter 3). The following two examples describe typical cases in which rights expressions occur in DRM:

Rights expressions are used within the DRM content consumption function (see Section 2.3.1). The digital content is rendered by the secure viewer according to earlier defined usage and access rights. These usage and access rights are formulated as rights expressions that either come with the content in a secure container or are received later on access demand. Thus, in this scenario rights expressions are exchanged within *one* DRM system, i.e. between DRM system components. They are interpreted and processed by the secure viewer. Rights expressions are used in IBM's DRM system, the Electronic Media Management System (EMMS) [IBM02] (see Section

---

[13]See: http://www.w3.org/
[14]See: http://www.educanext.org/

3.5). In EMMS, the rights expressions are packed into the secure containers (formerly *Cryptolope* technology [Kap96]). Microsoft's WMA format uses rights expressions formulated in a rights expression language to render various content types (music, video, etc.) with their Media Player. The Media Player functions as Microsoft's secure viewer. Nokia is currently developing a consumption technology [Nok01] for the mobile communications sector. Here, users receive content, such as images, ring tones, etc., in a secure container and have to purchase so called 'vouchers' (comprising rights expressions) to render the content on their mobile phone.

A number of providers of different digital and physical goods or services aim at offering their products via a common web portal and exploit the network effect (see Section 1.2). A typical example for this form of networking is Amazon.com[15] that started this strategy on a large scale very early and thus was able to realise first mover advantages [RvdV04]. Amazon, which formerly started as web portal for purchasing books, is now offering a large variety of products from digital cameras to apparel and toys. To technically implement this network the DRM systems of the providers have to formulate offers (i.e. rights expressions) comprising the description of the goods and their terms and conditions for Amazon. When offering a large number of different digital goods, with different conditions, such as music or video files, the offers can not be formulated manually. The offers therefore have to be available in a standardised format to be fed into the offer database of Amazon. This standardised format can potentially be a rights expression languages. In this example, the exchange of rights expressions occurs between DRM systems (content sellers and a content distributer) in the supply chain rather that between DRM system components.

The exchange of rights expressions occurs in all stages of the supply chain, i.e. between content creators, content packers, content managers, content publishers, content sellers, content distributers, and consumption service providers. Another application area of rights expressions are service level agreements (SLA) (see Section 4.9). Service level agreements coordinate the access of customers to certain services, e.g. a slot of the physical mobile telephone network to a mobile network operator (e.g. T-Mobile). Generally, the main field of application for rights expressions formulated in a REL is the exchange of rights information between interoperating DRM systems or their components, independent of the logical construct the rights expressions represent (contract, offer, etc.) or the application (secure viewer, accounting software, etc.) they serve.

---

[15]See: http://www.amazon.com/

In order to use a REL for the exchange of rights expressions between DRM systems respectively between their components, at least two technical tools have to be available [GSZ03]:

- *Rights expression generator.* This component supports the user in writing rights expressions, e.g. in the form of a web-based graphical user interface (GUI) that helps content providers to create offers. In that case, a rights expression is generated that expresses a digital good, the usage and access rights that the provider intends to grant to it, and the respective terms and conditions under which the rights are granted.

- *Rights expression interpreter.* A rights expression is not processable without an interpreter which is able to read and interpret the rights expression. For example, a secure viewer in charge of handling a secure container must be able to interpret the license that applies to content in order to grant access to and render the content accordingly. This type of processing is also called rights enforcement (see Section 4.7).

In the sample process described in 2.3.3 the generating tool is part of the DRM platform and supports the activity metadata enrichment of the DRM function *Content Preparation*. The interpreter tool, in turn, is part of the client software (or secure viewer) supporting the DRM function *Content Consumption*.

Chapter 3

Rights Expression Languages (RELs)

This chapter provides an insight to the field of rights expression languages (RELs). It justifies the application of rights expression languages in rights expressions respectively electronic contracts in today's DRM systems and addresses the requirements which have to be met by these languages (see Section 3.2). Section 3.3 addresses the language syntax and vocabulary of RELs. Standardisation is a critical success factor for RELs. If a REL has been accepted from or is supported by a standardising body, such as the W3C[1] or the Open Mobile Alliance[2], it is likely to become globally accepted respectively applied (as is exemplified with XML or RDF). Therefore, Section 3.4 introduces REL initiatives and their background, as well as languages that have already been accepted of standard–setting bodies. Practical examples (XML instances) of rights languages are given for the RELs ODRL and XrML. Finally, the chapter provides a short survey of the current market situation and trends in the field of DRM middleware and implementations using RELs (see Section 3.5).

3.1 Definition of Terms

Rights Expression Language: A rights expression language is a means of expressing usage and access rights of parties to assets. Rights

---
[1] See: http://www.w3.org
[2] See: http://www.openmobilealliance.org/

expression languages provide a syntax and semantics that are sufficiently rich to formulate rights expressions for digital publications, audio and video files, images, games, software, and other digital or physical goods, including pricing models as well as terms and conditions, regardless of whether a monetary consideration is part of the transaction. Consequently, rights expression languages provide *metadata framework* for the expression of rights.

REL instance or rights expression: Every document that is formulated in a rights expression language shall be defined as 'REL instance'. REL instances are exchanged between DRM systems respectively DRM system components and serve as interface between them. After being exchanged REL instances are further processed in applications, such as access control, accounting, etc. Sometimes the term *rights expression* is used as a synonym for REL instance. Depending on their content, rights expressions can represent different semantic constructs, e.g. licenses, digital tickets, or contracts (see Chapter 4, and in particular Section 4.3.1). A REL can be used to formulate "simple" rights expressions, such as "party X has the permission to play the resource Y", as well as complex electronic contracts where all contracting parties, the traded resources or rights, and the terms and conditions are specified in great detail. The mandatory elements in REL instances are defined in the respective REL specification. The difference between a "simple" rights expression and an electronic contract lies within content and its semantics. A contract requires at least the following elements: parties (consumer and rightsholder), resource, and permission.

3.2 Requirements of RELs

In order to provide a means of expression rights of parties to assets, a REL has to fulfill several technical and conceptual requirements. One substantial technical requirement of RELs is machine readability. Documents are *machine readable*, if a computer is able to digitally record the document information. Various techniques meet this requirement, e.g. even a newspaper article is machine readable after scanning the article and processing the image with an OCR (optical character recognition) software. OCR software, however, is not 100 percent reliable in terms of correct symbol recognition. Most of today's RELs are developed for the serialisation in XML, allowing for a formal representation of electronic contracts. Reading XML documents by a machine is more reliable than reading scanned documents and thus XML qualifies as an exchange format for rights expressions. XML is described in more detail in Section 3.4.1.

In Section 2.3.1 the function Content Consumption is described with the following sample DRM activities: handling the access request of the consumer, authenticating and identifying the consumer, identifying the resource that is subject to the access request, decrypting and decompressing the digital good(s), granting or denying access to the digital goods according to the license, and rendering the digital goods according to the granted permissions in the license. A number of REL requirements can be derived from these sample activities. In order to provide the relevant metadata, the REL should support

- *identification mechanisms:* unique ids are needed for the identification of parties (e.g. x500 [IT93a]) and resources (e.g. DOI [Nat00], ISBN [ISO92], ISSN [ISO98], etc.).

- *the definition of usage and access rights:* usage and access rights are e.g. play, print, copy, etc.

- *the definition of permission and duties:* permissions and duties, such as 'play Ebook no. 12356' or 'pay €100.00' are ⟨operation, object⟩ pairs where operation is an action (play) that may be or has to be performed on a certain resource.

- *the definition of constraints:* constraints are needed to narrow duties or usage and access rights in time, location, device, etc.

- *the articulation of roles:* some security mechanisms grant or deny access to resources depending the user's role rather than his/her identification number. Therefore, RELs also have to provide the possibility to express user roles, respectively express usage and access rights for roles.

- *the definition of technical details:* this supports the handling of decryption algorithms, viewers, etc.

- *workflow data:* this supports the course of the DRM process.

This informal enumeration does not represent a complete list of requirements for a REL. The Moving Picture Experts Group (MPEG) has specified the requirements for a rights expression language and its rights data dictionary in detail for the multimedia domain [Bor02]. The MPEG requirements comprise those listed above and a large number of additional ones, such as concepts for content aggregation, the sequencing of elements, etc. The work of Neal et al. [NCL+03] deals with the requirements for a special *Business Contract Language*. The definition of time constraints is addressed in detail, but the sequencing of operations (e.g. order, deliver, pay), and general constraints are only shortly sketched. The MPEG documents provide a comprehensive list of requirements and will be used as reference list for REL requirement in this thesis. The requirements of a rights language vary depending on their application field and scope, consequently RELs should be open and extensible.

3.3 Characteristics of RELs

Two basic factors in a language are its *syntax* and *lexis* (or vocabulary). The vocabulary of a language includes words that are created from permitted symbols (e.g. letters, numbers, and symbols). The syntax applies to the language vocabulary with which syntactically valid sentences can be formulated. Another crucial issue in language analysis is the field of *semantics*. The term semantics refers to the study of *meaning* as encoded in language [Wid96]. Syntax, lexis and the semantics of RELs are usually defined in a document called language specification [Ian02b, Con00, DWW03, Oct02]. The vocabulary of rights expression languages is sometimes referred to as *rights data dictionary (RDD)*) [Rig02, BR02]. With this specification an offer, contract, or other rights expression construct can be formulated.

Rights expression languages facilitate the interoperation of DRM systems and their components. They allow expressing rights information in a static format, i.e. putting down certain rights and conditions at a certain time/state. This static format provides

- *The separation of space:* Geographically distributed systems deploying different DRM technologies obtain means to communicate and interoperate.

- *The separation of time:* A contract is expressed at a certain state. At any time, business partners can re–read the contract and its conditions and verify its validity.

Rights expression languages have the potential to express aggregated rights information, i.e. RELs can phrase a good deal of rights information in a reduced representation. For example, permissions can be granted to groups of people (e.g. all students may access the script of informatics) or permissions can be granted to a type of resource that includes a large number of actual resources (e.g. user sguth may access all learning resources of the Department of Information Systems). Of course the software service that is processing such rights expression must be able to map the aggregated information to real objects.

3.3.1 REL Syntax

The basic elements in every REL syntax are permissions, resources and parties; the terminology for these three basic elements vary in each REL.

- *Permissions* are certain use or access rights to digital or physical goods or services. For the purpose of this thesis, *permission* is defined as operation–object pair. An operation is a certain action that can be performed on goods or services (objects), such as print, play, use, etc. Example for a permission is $\langle print, test.pdf \rangle$. Permissions can be specified in more detail by constraints. Constraints describe terms and conditions that have to be fulfilled before an operation is granted respectively serve to narrow the granted operation by time, location, individual, etc.

- The *resources* (or objects) represent the digital goods or services which the operations refer to. Resources have to be described by a non–ambiguous identifier such as DOI [Nat00].

- The *party* element represents any kind of party, i.e. a legal entity or a physical person which has a relationship (e.g. owns, controls, has permission to) to a digital product or service. In contracts, the party elements predominantly represent the people who enter into the contract. Examples of parties are the rights holder, the creator, the content provider, the consumer, the administrator, the beneficiary and the like.

Starting from these basic elements, each REL contains additional concepts for expressing containers, sequences, royalties, constraints, etc., and their relationships in more detail. Unfortunately, the REL community has not yet agreed on a general terminology for the basic REL elements. In one language the operation *play* is called a *right*, whereas in the other language *play* is indicated as *permission*. Following the terminology of this thesis a permission is an operation–object pair. Due to the longer history and available research, this definition has been adopted from the access control community.

3.3.2 Rights Data Dictionary (RDD)

The rights data dictionary (vocabulary) of a REL defines the words that are permitted in REL instances and their semantics. For example, in a REL instance the terms print, play, or view may be used as operations and the terms time, location, and individual may be used to constraint permissions. The table below shows an extract from the ODRL Data Dictionary in which several operation elements are defined. Similar vocabulary definitions exist for other ODRL syntax elements, such as ODRL constraints, and the ODRL context element. Each term is usually defined by a name, an identifier, and a description. The description denotes the informal semantics to a certain term. The ODRL Data Dictionary is compliant to the ISO–11179 standard which provides naming and identification principles for data elements [ISO95].

Name	Identifier	Description	Comment
Play	play	The act of rendering the asset in audio/video form.	...
Print	print	The act of rendering the asset on paper or hard copy form.	...
Execute	execute	The act of executing the asset.	...
...			

Other rights expression languages define their vocabulary in the same way (e.g. <indecs>rdd [RB99]) or similarly. XrML [Con00], for example, defines the informal, textual semantics of each lexical item in a small paragraph that also includes the respective extract of the XML schema defining the term. Additionally, the paragraph exemplifies the term's usage, its relations to other terms, and exceptions. Therefore, the XrML specification version 2.0 is very complex. In contrast XrML, the clear RDD definition (due to the application of the ISO–11179 standard) in ODRL is more comprehensive. All languages introduced in this chapter allow for an extension of the RDD via XML subschemata.

RELs are often more powerful than the DRM system requires. Therefore, the rights expression language is usually adapted to the specific implementation and domain, i.e. a subset of the vocabulary or only a restricted syntax is used. For example, in the Colis[3] project only a subset of access rights occurs in rights expressions [Ian03c]. For the purpose of this thesis, such adaptations are called *REL application policies*. Apart from defining the vocabulary subset such policies can also state the permitted identification schemes in instances (e.g. DOI, ISSN) or the depth of nested rights expressions. Application–specific rights expression generators and interpreters have to implement these policies.

3.4 Existing Rights Expression Languages and Initiatives

In this chapter, the most commonly used specifications in the field of rights expression languages are introduced. The field is still evolving, but the standards mentioned below have managed to prevail.

3.4.1 Open Digital Rights Language (ODRL)

The Open Digital Rights Language (ODRL) [Ian02b] is being developed by the ODRL initiative[4]. The ODRL initiative is an international effort which aims at developing an open REL standard. In the spirit of the open source community, ODRL is freely available. It was recently accepted by the Open Mobile Alliance (OMA)[5] as the standard REL for mobile content. The OMA aims at facilitating global user adoption of mobile data services. Therefore, OMA is developing specifications that ensure service interoperability across devices, geographies, service providers, operators, and networks, while allowing businesses to compete through innovation and differentiation. The latest version of the ODRL specification (version 1.1) has been co–published by W3C (as a W3C Note). The OpenIPMP Open Source Rights Management Project[6] has just released the first version of their DRM software that utilises ODRL for formulating rights expressions. The ODRL initiative does not have a focus on some particular application domain. However, ODRL is well accepted in the telecommunication domain

---

[3]See: http://www.colis.mq.edu.au/
[4]See: http://www.odrl.net/
[5]See: http://www.openmobilealliance.org/
[6]See: http://www.openipmp.com/

(through adoption of ODRL by OMA), in the educational domain, e.g. in the COLIS[7] project, the *le@rning federation*[8], the Open Archives Initiative [Bir01], and of the Dublin Core initiative [PJ02]. As ODRL is an open-source project, it is likely that the further development of ODRL will be research driven, i.e. researchers from all over the world will participate in future versions of ODRL. The release of ODRL version 2.0 is scheduled for the end of 2004. As a part of this thesis an interpreter for ODRL has been implemented. Therefore, ODRL will be discussed in more detail.

The following paragraphs present a REL syntax example of the straight-forward concept of ODRL. Referring to the earlier definitions, an *operation* in ODRL, e.g. play or print, is called *permission* (!), resources are indicated in ODRL as *assets*, and parties are also called *parties* in ODRL. The root element in ODRL is the *rights* element (see Figure 3.1), which represents one rights expression (e.g. a license, contract, etc.). The rights element can contain the rights expression itself with the *party*, *asset* and *permissions* elements or, alternatively, it can use the *offer/agreement* element to indicate semantically that a given rights expression is an offer or agreement. ODRL offers three different types of constraints: requirements, constraints, and conditions.

- If a *requirement* is defined in ODRL, the permission it is related to may not be granted prior to the fulfillment of this requirement. Payments are the most common requirements of ODRL.

- The *constraint* element in ODRL is designed to narrow ODRL permissions. For example, the permission play can be constrained to five times, by using the bound constraint *count*. ODRL provides for user–, device– bound–, temporal–, aspect–, target–, and rights constraints.

- An ODRL *condition* is oppositional to ODRL requirements. Once a condition is fulfilled, the respective permission is revoked.

If the ODRL rights expression includes a digital signature, the corresponding *XML Signature* [BBF+02] conforming information can be integrated into the document by means of the ODRL *signature* element. The ODRL syntax allows the addition of XML elements that are compliant with the XML Signature namespace. Figure 3.1 illustrates the elements discussed, which are merely a subset of the ODRL syntax. The entire current *foundation* model of ODRL is shown in Appendix A 9.1 (for a full

---
[7]See: http://www.colis.mq.edu.au/
[8]See: http://www.thelearningfederation.edu.au/

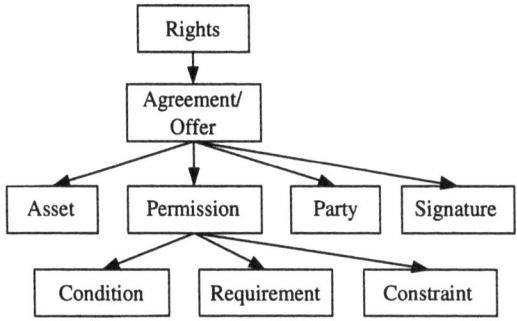

Figure 3.1: A subset of the ODRL language syntax

description of the concept, please refer to Iannella [Ian02b]). All ODRL elements can be further described by means of an ID, name, etc. with the help of the *context* element (not shown in Figure 3.1). Note that the ODRL model is further developed on the basis of the data model presented in this thesis. Please find the latest development of the ODRL foundation model at the ODRL Initiative web site[9].

Excursus: XML as Language Definition Framework

Extensible Markup Language (XML) [BPSMM00] is a meta–language for the definition of application, respectively domain–specific markup languages. Consequently, XML is a means to define new markup languages for a certain domain, e.g. rights expressions. XML is a successor of the Standard Generalized Markup Language (SGML) [ISO86]. A new XML–based markup language defines XML elements and their XML types (e.g. complex type), as well as their structure, i.e. how these elements may be arranged in an instance of the language. XML documents have a well–defined document structure and are human as well as machine readable. XML plays an increasing role in the exchange of data on the web and elsewhere.

The code that defines new markup languages or document types in XML is called Document Type Definition (DTD) [BPSMM00] or XML Schema [TBMM01, BM01]. In contrast to DTDs, XML schemata provide rich datatyping capabilities for elements and attributes (e.g. String, Integer),

---

[9]See: http://www.odrl.net/

object-oriented design principles (e.g. inheritance), and namespaces. A document that is derived from a specific XML schema or DTD is called XML document or XML instance. An XML instance is *valid* if it conforms to that specific XML schema or DTD. In order to check whether an XML instance is valid (with respect to a specific XML schema or DTD) is is 'validated' against the respective XML schema or DTD (see the Example below). An XML instance is *well-formed* if it contains at least one element, if it has a unique opening and closing tag, if the tags are nested properly (i.e. there must be an opening and a closing tag that do not overlap), and if the attribute values are quoted[BPSMM00]. XML documents contain *one* root element; an XML root element and its nested elements below is sometimes referred to as *XML tree*.

All rights expression languages introduced in this chapter are defined in XML schema documents. For the better understanding of subsequent chapters, a small introduction to XML Schema is exemplified with ODRL. To make the example more comprehensive, the following code (see Figure 3.2) is a simplified subset of the XML Schema defining the Open Digital Rights Language (for the complete ODRL schema, see Appendix A)

As mentioned above, XML Schema documents define permitted elements, as well as their types and structure. In the example, a number of elements are defined, namely rights, offer, agreement, as well as asset, party, permission, and context and finally uid, date, and remark. Each of these elements have types; element types can either be simple or complex. Elements with simple types do not have further elements, and represent the 'leaves' of an XML tree, whereas complex element types comprise one or more further elements. The elements right (rightsType), offer, agreement (offerAgreeType), and context have complex element types. An element of the rightType may comprise offer and/or agreement elements.

To reduce complexity, within the definition of a complex type it can be referred to element definitions, e.g. as to offer and agreement in the rightsType definition. The elements offer and agreement are of the complex offerAgreeType, comprising the elements context, party, asset, and permission. Finally, the element context is defined as complex type, comprising the elements uid, date, and remark. The element uid, date, and remark , as well as the elements asset, party, and permission are of the simple type xsd:string and may therefore simply comprise words or sentences.

```xml
<?xml version="1.0" encoding="UTF-8"?>
<xsd:schema xmlns:xsd="http://www.w3.org/2001/XMLSchema"
            xmlns:o-ex="http://odrl.net/1.1/ODRL-EX"
            targetNamespace="http://odrl.net/1.1/ODRL-EX"
            elementFormDefault="qualified" attributeFormDefault="unqualified">

    <xsd:element name="rights" type="o-ex:rightsType">
        <xsd:annotation>
            <xsd:documentation>Root element of each ODRL RE</xsd:documentation>
        </xsd:annotation>
    </xsd:element>

    <xsd:element name="offer" type="o-ex:offerAgreeType"/>
    <xsd:element name="agreement" type="o-ex:offerAgreeType"/>

    <xsd:complexType name="rightsType">
      <xsd:choice minOccurs="0" maxOccurs="unbounded">
         <xsd:element ref="o-ex:offer" minOccurs="0" maxOccurs="unbounded"/>
         <xsd:element ref="o-ex:agreement" minOccurs="0" maxOccurs="unbounded"/>
      </xsd:choice>
    </xsd:complexType>

    <xsd:complexType name="offerAgreeType">
      <xsd:choice minOccurs="0" maxOccurs="unbounded">
         <xsd:element ref="o-ex:context" minOccurs="0" maxOccurs="unbounded"/>
         <xsd:element ref="o-ex:party" minOccurs="0" maxOccurs="unbounded"/>
         <xsd:element ref="o-ex:asset" minOccurs="0" maxOccurs="unbounded"/>
         <xsd:element ref="o-ex:permission" minOccurs="0" maxOccurs="unbounded"/>
      </xsd:choice>
    </xsd:complexType>

    <xsd:element name="asset" type="xsd:string"/>
    <xsd:element name="party" type="xsd:string"/>
    <xsd:element name="permission" type="xsd:string"/>

    <xsd:element name="context">
      <xsd:complexType>
         <xsd:choice minOccurs="0" maxOccurs="unbounded">
            <xsd:element ref="o-ex:uid" minOccurs="0" maxOccurs="unbounded"/>
            <xsd:element ref="o-ex:date" minOccurs="0" maxOccurs="unbounded"/>
            <xsd:element ref="o-ex:remark" minOccurs="0" maxOccurs="unbounded"/>
         </xsd:choice>
      </xsd:complexType>
    </xsd:element>

    <xsd:element name="uid" type="xsd:string"/>
    <xsd:element name="date" type="xsd:string"/>
    <xsd:element name="remark" type="xsd:string"/>
</xsd:schema>
```

Figure 3.2: A simplified subset of XML schema defining ODRL

```xml
<?xml version="1.0" encoding="UTF-8"?>
<rights xmlns="http://odrl.net/1.1/ODRL-EX"
        xmlns:xsi="http://www.w3.org/2001/XMLSchema-instance"
        xsi:schemaLocation="http://odrl.net/1.1/ODRL-EX
        http://wi.wu-wien.ac.at/Wer_sind_wir/Guth/schemas/odrl-simple.xsd">
    <agreement>
        <context>
            <uid>agreement #112233</uid>
            <date>12/31/03</date>
            <remark>This agreement was concluded in Vienna/Austria</remark>
        </context>
        <party>Susanne Guth</party>
        <permission>play</permission>
        <asset>Hit clip #999888</asset>
    </agreement>
</rights>
```

Figure 3.3: A valid language instance of the simplified ODRL schema

At the beginning of the XML schema, there is a definition of which character set can be used in instances (UTF-8), which namespace has been used to write the schema (http://www.w3.org/2001/XMLSchema), and which namespace the schema at hand provides (http://odrl.net/1.1/ODRL-EX). The XML schema also specifies that all elements in instances of this schema have to be assigned to a qualified namespace (elementFormDefault). The attribute attributeFormDefault denotes that only the globally declared attributes must be namespace qualified in instance documents; locally declared attributes are not namespace qualified. From the schema above a respective ODRL instance can be derived (see Figure 3.3). It is important to note that this is a simplified XML instance that is not compliant with ODRL version 1.1. The two XML schemata defining ODRL version 1.1 can be found in Chapter 9 (Appendix A). Stating rights information in an XML-based language provides flexibly, as language elements from other XML schemata can be integrated, i.e. XML-based RELs permit to reuse of description languages, such as the Learning Object Metadata (LOM) standard [IEE02] or Dublin Core [Dub01] or XML Signature [BBF+02] .

Several free open-source tools are available for the work with XML schemata and instances. Tools for processing XML documents are called XML parsers.

XML Parser. An XML parser, such as Expat[10] or Xerces[11], is a software tool that receives input in the form of XML markup tags and breaks them up into parts (for example, the nouns, verbs, and their attributes (or options)) that can then be managed by other software services, e.g. a language interpreter. Some parsers include XML validators, such as PyTREX[12] (for Python platform) or DSD Processor[13] (for Java platform). A validator checks if an XML document is valid in respect of a certain XML schema. A large number of other XML parsers can be found on the W3C web page for XML schema[14].

ODRL Example

ODRL version 1.1 includes two XML schemata: one that defines the language syntax and a second that defines the ODRL rights data dictionary. The XML schema defining the ODRL rights data dictionary is basically an extension of the XML schema defining the ODRL syntax. For example, in the syntax an element called permissionElement is defined. Via the XML mechanism *substitutionGroup* the data dictionary defines all terms that can be used as permissionElement, e.g. play, print, copy, etc. Both the XML schema of the ODRL syntax and rights data dictionary can be found in Appendix A. The following code is an example compliant to ODRL version 1.1, showing a contract for a video (disregarding XML namespace labels). ODRL uses XML attributes to assign additional information to the ODRL vocabulary (see "currency" of the amount tag).

The sample license shows a recording of a marketing lecture sold to the *Université Libre de Bruxelles* for the price of €*10.00* with the permission to *play* the video *five times*. The video stream's rights holder is the *Department of Information Systems at the Vienna University Economics and BA*. In this example, ids from the numbering system of the Universal Project[15] are used.

```
<rights>
 <agreement>
  <party>
```

---

[10] See: http://www.jclark.com/xml/
[11] See: http://xml.apache.org/xerces-c/
[12] See: http://pytrex.sourceforge.net/
[13] See: http://www.brics.dk/DSD/
[14] See: http://www.w3.org/XML/Schema
[15] See: http://www.ist-universal.org/

```
<context>
 <uid>urn:univ:us-wuw-deptIS </uid>
 <name>Department of IS, Vienna Univ. of Economics and BA</name>
</context>
<rightsholder/>
</party>
<asset>
 <context>
  <uid>urn:univ:lr-wuw-vid-1</uid>
  <name>Marketing strategies for Universal</name>
 </context>
</asset>
<party>
 <context>
  <uid>urn:univ:us-wuw-uniBrux</uid>
  <name>Université Libre de Bruxelles</name>
 </context>
</party>
<permission>
 <play>
  <requirement>
   <prepay>
    <amount currency=EUR>10.00</amount>
   </prepay>
  </requirement>
  <constraint>
   <count> 5 </count>
  </constraint>
 </play>
</permission>
</agreement>
</rights>
```

This example reflects the ODRL syntax illustrated in Figure 3.1. The basic elements within an agreement are *party, asset,* and *permission.* The party element occurs twice, for the consumer and for the rights holder. The rights holder is identified by the ⟨rightsholder⟩ element nested below a party element. The language specification defines that permissions on the same XML tree level as assets, refer to these assets (if no further references are specified). Likewise, assets and permissions are related to customers. The example above does not use a signature element. Constraints (such as ODRL requirements and ODRL constraints) are directly nested below ODRL permissions.

3.4.2 eXtensible rights Markup Language (XrML)

The eXtensible rights markup language (XrML) [Con00] is a rights expression language developed by ContentGuard[16], a spin–off of Xerox in cooperation with Microsoft. The language XrML itself is free of charge, but ContentGuard holds a US patent on the usage of rights expression languages in general. ContentGuard claims that its patents pertain "*to the distribution of digital works and to any rights language*". The interpretation and the consequences of this patent are not clear and are often discussed in DRM/ REL–related communities. After working through various online resources, such as web pages and news groups, and personally discussing this issue with ContentGuard employees, I have come to the conclusion that usage rights for industrial/commercial use in the US of XrML or any other rights expression language need to be licensed with ContentGuard.

ContentGuard aims at applying XrML in DRM systems that focus on the commercial exchange of digital goods. Therefore, the standard vocabulary of XrML is designed to express a large number of pricing models. Today XrML is used in Microsoft products (see Section 3.5). Most likely, the further development of XrML will be industry–driven. Is is likely that Microsoft will be able to use its impact to the further development of XrML and the above named patents as competitive advantage in the market for DRM system solutions.

XrML Example

XrML is defined by three XML schemata: the XrML core schema, the XrML standard extension (sx) schema and the XrML content extension (cx) schema. The following example includes XML namespace information, which is necessary for the validation of elements from different namespaces. Just as ODRL, XrML envisages the use of XML Signature [BBF+02] to specify the identity of the contracting parties. The example below shows an XrML instance which reuses elements of the XML Signature namespace.

The "license"–tag is the root element of an XrML instance, resource and party are referred to as the "resource" and "principal" in the basic syntax of XrML. "Grant" includes the actual rights expression. Operations are expressed as "rights" and constraints as "conditions". The XrML–compliant representations of resource and consumer party are "digital work" and "key-

---
[16]See: http://www.contentguard.com/

Holder." The XrML vocabulary contains "print" and "validityInterval" as an operation and condition. The XrML license below grants the owner of the x509 certificate the use of *someResource* until the *end of 2005*.

```xml
<?xml version="1.0" encoding="UTF-8"?>
<license xmlns="http://www.xrml.org/schema/2001/11/xrml2core"
         xmlns:sx="http://www.xrml.org/schema/2001/11/xrml2sx"
         xmlns:dsig="http://www.w3.org/2000/09/xmldsig#"
         xmlns:xsi="http://www.w3.org/2001/XMLSchema-instance"
         xmlns:cx="http://www.xrml.org/schema/2001/11/xrml2cx"
         xsi:schemaLocation="http://www.xrml.org/schema/2001/11/xrml2cx
  <grant>                                       ..\schemata\xrml2cx.xsd">
    <keyHolder>
      <info>
        <dsig:x509Data>
          <dsig:X509IssuerSerial>
            <dsig:X509IssuerName>CN=Guth Susanne,
                 OU=Dept. of Information Systems,
                 O=Vienna University of BA, L=Vienna,
                 ST=Vienna, C=Austria
            </dsig:X509IssuerName>
            <dsig:X509SerialNumber>12345678</dsig:X509SerialNumber>
          </dsig:X509IssuerSerial>
          <dsig:X509Certificate>MIIEODCCA6GgAwIBAgIBEDANBgkqhki...
                 ...Zos6NAm8m6UQBA== </dsig:X509Certificate>
        </dsig:x509Data>
      </info>
    </keyHolder>
    <cx:print/>
    <cx:digitalWork>
      <cx:locator>
        <nonSecureIndirect URI="http://www.wu-wien.ac.at/someResource"/>
      </cx:locator>
    </cx:digitalWork>
    <validityInterval>
      <notAfter>2005-12-24T23:59:59</notAfter>
    </validityInterval>
  </grant>
</license>
```

In XrML the elements keyHolder, operation (e.g. print), digitalWork, and constraint (e.g. validityInterval) are positioned on the same XML tree level.

Therefore, like in ODRL, the keyHolder is related to the operation *print* and print refers to *someResource*. Likewise, the constraint is referred to the operation *print*. The latter semantics is different from ODRL, where constraints are nested directly below the respective operation. Thus, in XrML the constraints are related to the operation *and* the resource (digitalWork), whereas in ODRL the constraints are related only to the operation. This fact illustrates one of a probably large number of differences in the syntaxes of the two languages. Such syntax differences between the two RELs are hard to identify as unfortunately, until now no formal semantics has been developed, neither for ODRL nor for XrML. XrML seems to be focused more on the commercial aspect of a rights expression language, i.e. on expression licenses that a sold and issued by a DRM platform and bought respectively executed by consumers. In contrast to ODRL, XrML does not seem to focus on the formulation of contracts as in XrML not contract parties are specified but issuers and keyholders.

3.4.3 MPEG 21

The Moving Picture Experts Group (MPEG)[17] is the ISO/IEC working group in charge of developing standards for the coded representation of digital audio and video. Among other standards, MPEG is working on MPEG 21 with the intention to develop a standardised multimedia framework. Parts 5 [DWW03] and 6 [BR02] of the MPEG 21 standard specify a REL respectively RDD suitable for such a framework. After defining the *requirements for RELs and RDDs* [Bor02], MPEG issued a call for contributions to select one REL and one RDD as a basis for future development. XrML version 2.0 has been accepted as basis for the development of a future MPEG 21 rights expression language, and the data dictionary from the <indecs>initiative has been accepted as the basis for the future Part 6 (RDD) of the MPEG 21 standard.

The <indecs>2rdd Project

The <indecs>2rdd project is based on the <indecs> project, which defined a framework for interoperable metadata in content–based e–commerce and is now hosted by the DRM consulting company Rightscom[18]. In contrast to ODRL and XrML, the project does not provide a syntax, but focuses

---

[17]See: http://mpeg.telecomitalialab.com/
[18]See: http://www.rightscom.com/

exclusively on defining a rights data dictionary. Thus, <indecs>2rdd is not a rights language but can be adapted from a REL as RDD. The rights data dictionary of the <indecs>2rdd project aims at providing a more sophisticated RDD than XrML and ODRL do, and therefore introduces a rights ontology which supports interoperability between the various RELs. The <index>2rdd project is currently working on the shaping of its RDD according to the requirements of MPEG 21.

3.4.4 LicenseScript

LicenseScript [CCL+03] is a rights expression language that is not defined in an XML schema. LicenseScript is a multi-set rewriting/logic-based language for expressing dynamic conditions of use of digital assets such as music, video or private data. LicenseScript differs from the other DRM languages in that it does not express a certain state, such as an XML contract, which states an agreement at a certain date, but it tracks the development of a rights expression from its issue to its consumption. This REL does not intend to provide a rights expression exchange format between differently designed DRM components or systems. Therefore, each system that aims at using LicenseScript has to use the implementation of the LicenseScript interpreter. Although this characteristic is reducing semantic errors when interpreting rights expressions, it also restricts the usage spectrum of LicenseScript. Each rights expression in LicenseScript is a small Prolog program. Therefore, the question arises if LicenseScript is a rights expression language. Another approach of logic-based rights expression languages has been discussed in [Sza02].

3.5 Current Market Situation and Trends

This section examines the application of rights expression languages in the current DRM systems market. The leading developers of DRM middleware are IBM, Adobe, Real Networks and Microsoft. Real Networks, however, is currently not using any of the introduced RELs in their products.

- IBM has developed a product called the Electronic Media Management System (EMMS)[19], which currently deploys a proprietary rights expression language influenced by ODRL. EMMS supports a variety of media formats. IBM is working in close cooperation with Nokia to develop solutions for the mobile communications sector [Nok01]. Nokia

---
[19]See: http://www.ibm.com/software/data/emms/

has just released a new version of their content publishing toolkit that provides a content creation that meets the requirements of OMA (OMA uses ODRL, see Section 3.4.1) and enables deployment of content and rights to mobile handsets.

- Microsoft has implemented XrML in its Windows Media™ Rights Manager. This software provides a means of packaging content and specifying usage and access rights formulated in XrML. The output of this tool is a file in the Windows Media format (WMA). XrML instances can be interpreted and processed, i.e. enforced, by the Windows Media Player.

- Adobe offers DRM solutions for the exchange of documents including e-Books in PDF. The documents are created with the Adobe Content Server software and can be interpreted and enforced with the corresponding reader, which offers the proprietary functionality of a secure viewer. Adobe is a supporter of the ODRL initiative and a DRM player which will potentially use ODRL in future products. Today, a proprietary format to express rights is used in Adobe's software.

Based on this middleware, some implementations have already appeared on the Internet. One of the first music subscription services, PressPlay[20], uses the Microsoft solution and thus works with XrML. MusicNet[21] is a digital music service based on Real Networks' technology. The M-Stage Mobile Music Service[22] is a product on the Japanese mobile-commerce market hosted by NTT DoCoMo, based on IBM's EMMS technology. Apart from the market leaders, there are also other projects which have implemented rights languages, such as the "Collaborative Online Learning & Information Services (COLIS)[23] project, which uses ODRL.

One good source of online information on RELs is the XML coverpages of OASIS' *The XML Coverpages*[24]. Another online source for DRM news (currently free of charge) is DRM Watch[25], which has become a commercially run platform for DRM content.

---

[20]See: http://www.pressplay.com/
[21]See: http://www.musicnet.com/
[22]See: http://www.nttdocomo.co.jp/p_s/mstage/music/
[23]See: http://www.colis.mq.edu.au/
[24]See: http://xml.coverpages.com/drm.html
[25]See: http://www.giantstepsmts.com/drmwatch.htm

Chapter 4
Electronic Contracts

This thesis deals with the exchange and processing of rights expressions. Electronic contracts are rights expression with a particular semantic meaning and have importance in terms of their legal effect. As this thesis especially focuses on processing electronic contracts, this chapter addresses the particularities when dealing with electronic contracts.

As soon as parties agree to exchange digital or physical goods, a contract is concluded.

> A *contract* is an agreement of two or more parties, i.e. a two or multilateral declaration of intent on the exchange of rights to goods or services under certain terms and conditions. The memorandum of an agreement is informal, i.e. can be stated verbally, or in writing, etc.

The rapidly growing interest in purchasing goods (e.g. music files, e-books, videos, or e–learning content) via the Internet is therefore accompanied by an increasing demand for contracts that are concluded via the Internet. Platforms that offer the exchange, respectively the purchase of goods or services are e.g. Amazon[1], eBay[2], or iTunes[3]. Every time customers desire to purchase goods and services offered via the Internet, they usually declare their intent to do so via a *click-through-agreement* [Ame01], resulting in a contract between seller and buyer. Parties that participate in

---
[1]See: http://www.amazon.com/
[2]See: http://www.ebay.com/
[3]See: http://www.itunes.com/

e–commerce are usually not at the same physical location. *Electronic contracts* support the conclusion of contracts between dislocated parties. For the purposes of this thesis, an electronic contract is defined as follows:

> A *digital/electronic contract* is an agreement of two or more parties, on the exchange of rights to (digital) goods or services under certain terms and conditions. The memorandum of an electronic contract is digital and can be transmitted via an electronic network.

In the European Union the declaration of intent in electronic contracts is legally binding if it has been stated in the form of an electronic signature [Eur99]. Due to their digital format, electronic contracts have the potential to be electronically processable. The memorandum of an electronic contract in a well–structured, standardised format increases its processability. A way to structure and standardise electronic contracts is the formulation in a rights expression language (REL) (see Chapter 3).

The remainder of this chapter is structured as follows: Section 4.1 introduces the contract life cycle and its four basic states *offer placement, contract conclusion, execution of contracts,* and *contract archiving*. Section 4.2 addresses the states and state transitions for electronic contracts. Section 4.3 gives a detailed insight into the execution of rights that result from contracts and their processing. At modelling level, an electronic contract can be seen as a composition of different contract objects with various attributes. In other words, an electronic contract aggregates a number of interrelated objects. The core objects, additional application–specific contract objects, and their attributes are addressed in Section 4.4 of this chapter.

The Section 4.5 is concerned with the application–specific generation of electronic contracts. First, various usage scenarios for the application of electronic contracts (e.g. access control, accounting) are identified (see Section 4.4.2) and shortly described. Then, in an example, the required contract objects and their attributes are derived for the usage scenario access control. In the last subsection, Section 4.5.3, a basic process for the tailoring of electronic contracts is proposed.

Section 4.6 addresses the pragmatics of electronic contracts, i.e. their processing in software services. For this purpose the generic Contract Schema (CoSa) in introduced that is an abstraction layer of rights expressions. Within one application or domain all rights expressions are mapped to one contract schema. The application programming interface of CoSa then allows a uniform querying of the contract information within that application respectively domain no matter what underlying representation the contract has

In Section 4.7 the enforceability of electronic contracts is addressed. Section 4.8 covers management issues, such as contract validity, digital signatures, identification of contract content, when dealing with electronic contracts. The chapter closes with various related projects and approaches in the field of electronic contracts.

4.1 Contract Life Cycle

The contract life cycle defines the different phases that are undergone by electronic contracts. The contract phases directly affect the required contract characteristics, such as the contract content, and also influence management issues with respect to electronic contracts. The contract phases can be considered from a technical and a legal view point. The legal phases are outlined in the *contract life cycle* and the technical phases are addressed in the *contract states*.

Figure 4.1 shows a simple contract life cycle from a non–technical standpoint. The phases were derived from considering the German Civil Code. The contract life cycle has four different phases: *the offer placement, the offer confirmation, the contract fulfillment, and the contract archiving.* The following paragraphs describe these four phases in more detail.

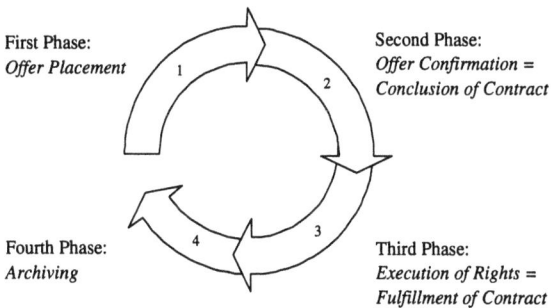

Figure 4.1: A simple contract life cycle with four phases

1. *Offer Placement.* A contract results from two (or more) declarations of intent about the exchange of rights to goods or services. A signed electronic offer is legally considered as a declaration of intent. Content owners (or rights holders) *offer* their goods to the consumers on certain terms and conditions. These terms and conditions describe the permissions and duties of the contracting parties (i.e. the consumer and the content owner).

2. *Offer Confirmation / Conclusion of Contract.* At this stage, the consumer confirms the electronic offer made by the content provider. The consumer does so by signing the offer (second declaration of intent) and thereby accepting the terms and conditions; this results in an *electronic contract.* Note that the contract conclusion is usually preceded by negotiations between content owners and consumers. In other words, a contract can only be concluded if the contracting parties have reached an agreement on the relevant terms and conditions. Sometimes, several new offers are placed until an agreement is reached. Every new offer may include significant modifications to the contract's original terms and conditions. In order to be legally *valid,* a contract has to be signed by all contracting parties.

3. *Fulfillment of Contract/Execution of Rights.* In this phase, the contract "fulfillment" takes place, i.e. the contract parties exercise their rights and fulfill their duties under the corresponding conditions. The chronological sequence of these actions can be specified in the contract (e.g. payment in advance). Once all rights have been exercised and all duties have been fulfilled, the contract is *completed.*

4. *Archiving of Contract.* After completion, each contract is saved in a permanent archive. However, the statutory period for which a particular contract has to remain archived depends on the type of contract (e.g. contract of sale, last will, etc.) and on local law.

4.2 Contract States

The various contract states and state transitions describe the process that electronic contracts undergo to move from one phase in the contract life cycle to the next. Figure 4.2 depicts a state chart diagram with the basic states and state transitions of electronic contracts. State transitions are specified by the required *event,* the *[condition]* and the respective */action.*

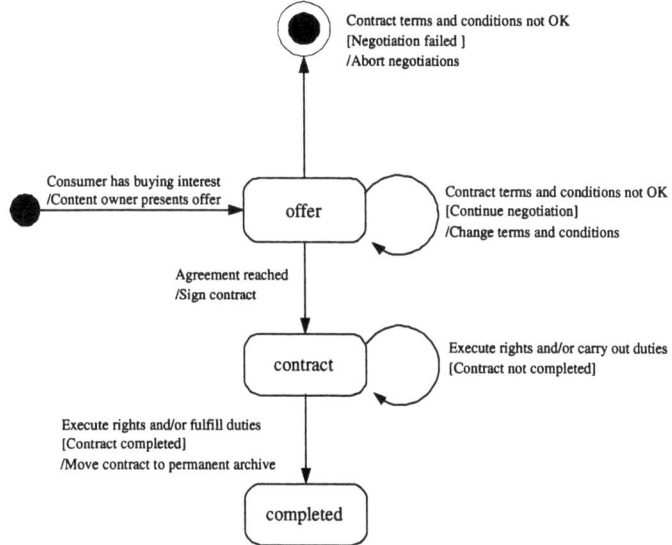

Figure 4.2: Basic states and state transitions of electronic contracts

Once a consumer has expressed interest for buying digital goods or services, the owner of the corresponding content places an *electronic offer* and thereby begins the negotiations. In general, the consumer may reject the offer, demand modifications to the offer, or accept the offer. If the consumer rejects the offer, because of unfavorable terms and conditions, and s/he does not wish to continue negotiation, then the negotiations have failed and are consequently aborted. If the consumer rejects the offer, but is willing to negotiate with the content owner, then the terms and conditions can be modified. Subsequently the consumer checks terms and conditions again. [GSSG00] presents the negotiation process with additional cases (e.g. an offer is not answered) that lead to the states in Figure 4.2. If an agreement is reached, both parties (consumer and content owner) will sign the contract. After the conclusion of contract the contract in turn becomes *valid*. Subsequently, both parties can execute the rights and/or have to fulfill the duties specified in the contract. Once all rights have been consumed and all duties have been carried out, the contract is fulfilled and then moved to a permanent archive. These basic states and state transitions can be extended and customised for software services that support this process.

4.3 Execution of Rights

This section deals with options for the execution of rights derived from electronic contracts. Rights are exercised in the *'fulfillment of contract'*-phase of the contract life cycle (see Section 4.1). The execution of rights is of particular interest for this thesis as it addresses the process of extracting rights information from contracts and forward it to other software services, such as access control. A detailed technical consideration of this process can be found in Section 7.2.

For the purpose of explaining the *'execution of rights'*-process in detail, in the first subsection the term *electronic tickets* will be introduced and distinguished from electronic contracts. Tickets can be derived from electronic contracts and redeemed as detached rights. To fully understand electronic tickets, Subsection 4.3.2 introduces tickets characteristics and some research that has been done in this field.

The term *contract right* in connection with the *execution of rights* has to be further explained: note that in general the contract duties of the content provider are the contract rights of the consumer, and vice versa. Therefore, instead of defining contract rights and contract duties separately, they can be expressed as contract rights only. For example, a contract has been concluded which states that the consumer has the *duty* to pay a certain amount in order to receive the *right* to visit a concert. This contract finally results in two rights, 1. the right of a customer to attend a concert and 2. the right of the concert promoter to collect the corresponding entrance fee from this particular customer. Consequently, a *contract right* is a triple that comprises an operation (e.g. collect, or access) which has been granted to one of the contract parties (e.g. consumer, provider, beneficiary) and which may be performed on certain objects (e.g. money, respectively digital goods or services) under certain terms and conditions. In contrast to this, a *permission* is defined as a pair that comprises simply an operation that may be performed on a certain object. An example for a permission is the pair (play, music track) (see Chapter 3). Figure 4.3 is an illustration of the two terms *contract right* and *permission*.

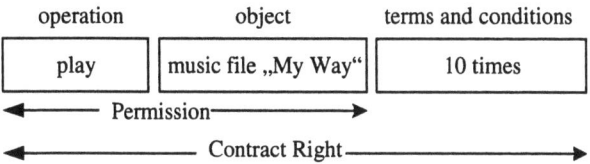

Figure 4.3: *Contract right* versus *permissions*

4.3.1 Electronic Contracts, Electronic Tickets, and Licenses

In this section, a definition of electronic tickets will be provided and the similarities and differences between electronic contracts and electronic tickets will be identified. In the definition of electronic contracts at the beginning of this chapter, these contracts comprise information about rights of contracting parties to goods or services (the object of the agreement), and their terms and conditions. Permissions are permitted operations to certain objects. Electronic tickets are defined as follows:

> A digital/electronic ticket is the option to consume a permission under certain terms and conditions.

Electronic tickets are sometimes called *voucher* [Nok01]. Electronic contracts and tickets both contain rights expressions pertaining to digital goods or services. However, owning a ticket is different from being party of an electronic contract. Among other things, a contract specifies several rights exchanged between the contracting parties, while an electronic ticket describes an excerpt from a contract, namely one (or more) *contract right(s)* which can be executed. At the beginning of this section, we mentioned that both contract rights *and* contract duties can be expressed in the form of rights. Consequently, each contract right specified in a contract can be extracted and formulated as an electronic ticket. In one frequently encountered situation, two parties conclude an electronic contract in which one party receives the right to consume specific digital goods, while the other party receives the right to collect money for these goods. When fulfilling a contract, its rights can thus be transformed into tickets which may be executed independently of each other.

For example, a rock concert promoter sells two tickets for admission to a person who is planning to see the concert with a friend. The right to

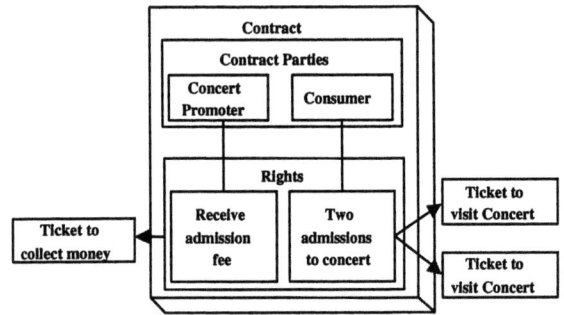

Figure 4.4: Contracts and tickets – an example

attend the concert with a friend (the right of the consumer) as well as the right to receive the concert entrance fee (the right of the concert promoter) can be extracted from the contract and formulated as stand-alone digital tickets. The concert promoter thus owns a ticket that allows him/her to collect money from the consumer, and in return the consumer receives two admission tickets to the concert (see Figure 4.4).

A *license* is a specific type of contract right, respectively a specific type of electronic ticket. A license grants usage rights to intellectual property, technical know–how or technical inventions [Sch03]. Consequently, a license is a ticket for the usage of intellectual property, technical know–how, or technical inventions. Licenses are traded in *license agreements.*

The chronological sequence in which the tickets are executed (or redeemed) can be specified in the electronic contract. Concert tickets, for example, usually have to be paid for in advance. Once the concert promoter has received the entrance fees, he issues the concert tickets to the consumer, thereby granting the right(s) to attend a particular concert.

There are a number of other approaches to facilitate the consumption of rights granted in an electronic contract. One interesting approach is described in the works of Fujimura et al. [FKT+99], Stefik [Ste97], and Rivest [Riv97] who define the term "digital ticket."

- Fujimura defines a digital ticket as "... a digital medium that guarantees certain rights of the owner and it includes software licenses, resource access tickets, event tickets, and plane tickets." This definition is closest to our approach but not considering the ticket as option and does not state what the rights are referred to.

- Stefik compares digital tickets with "... coupons found in a local paper that give discounts on the purchase of grocery products. Issued by a publisher, they correspond to prepayment or discounts for using works by the publisher."

- Rivest et al. have referred to digital tickets as a "... *means of payment*".

However, Stefik's and Rivest's definition do not correspond our definition given above. According to my definition in this thesis, tickets are part of the contract life cycle and occur in phase 3, "fulfillment of contracts / execution of rights" (see Section 4.1). As mentioned above, some rights specified in a contract might result in a ticket allowing one of the contracting parties to collect money from another contracting party. In this particular case, the digital ticket can be seen as "a means of payment." However, other types of contract rights, such as "attending a concert", results in a ticket issued by the concert promoter; in such cases the ticket serves as "a means of gaining admission." In my view, their definition of tickets as a "a means of payment" or "coupon for discount" is only one special case in which digital tickets can be applied.

4.3.2 Ticket-Driven Rights Execution

This section introduces different types of electronic tickets and their application fields. A digital ticket can be personalised (e.g. bound to a certain individual) or anonymous:

- *Anonymous.* A digital ticket is considered anonymous if only the issuer of the ticket can be identified, while the beneficiary remains anonymous. Since contracting parties can be identified by their digital signatures, an anonymous ticket has to be signed only by the issuer. In other words, a digital ticket must at least include the signature of the ticket issuer. The signature then serves as means for verifying the integrity of the digital ticket and authenticating the ticket issuer.

- *Personalised.* A digital ticket is personalised if it allows the identification of the consumer and the issuer. Anonymous tickets are used e.g. for concerts or bus fares, to name but two examples. In such a case, the identity of the person who executes a ticket is generally not important to the issuer. One possible use of personalised tickets is in airline ticketing, as airline companies are required to verify the identity of all passengers on a flight.

In many cases, all ticket information is available in the contract. It is reasonable ask: "*What are it sensible applications of electronic tickets*"? Examples might include the following:

- *Privacy*. If, for example, the contracting parties want to consume their rights anonymously, a ticket is a means of addressing this issue.

- *Efficiency*. As tickets are excerpts from contracts, they often comprise a smaller amount of data. The actual size of a digital document can be relevant for storage–restricted applications, for example transmitting electronic tickets to chip cards or SIM (subscriber identity module) cards.

- *Specific Ticket Information*. In some cases, consumption–relevant information (for example, the current download location of a digital resource) is not specified in the contract but has to be added to the ticket when issued.

In cases where a contract specifies that goods or services may be accessed a certain number of times, the ticket issuers can use two basic mechanisms to formulate a ticket. They can either issue a certain *number of equal tickets*, or *one ticket* that expires once all rights have been exercised. Depending on the intended use of the ticket and the technology used, both mechanisms can be appropriate. Issuing one ticket for each use may result in a large number of electronic tickets to be stored and managed. On the other hand, changing the number of "remaining uses" in the ticket after each use requires greater administration and security effort at runtime. Questions, such as "Who may edit the ticket?" arise in this case.

One proposal for a formal ticket language is XML Ticket [FNS99]. Every ticket formulated in XML Ticket can theoretically be expressed in a digital rights language, but not vice versa. XML Ticket only provides the syntax and semantics to specify a right which a ticket issuer grants to a (subsequent) ticket owner. The XML Ticket language is restricted to these two roles (issuer, owner) and provides no means of expressing other relevant information, such as payments methods, etc. Furthermore, the XML Ticket language is not subject to any current further development. Therefore, for the time being, I propose the formal expression of tickets in a digital rights language as well.

4.3.3 Hybrid Rights Execution

However, the fulfillment of a contract does not have to be regulated exclusively by tickets. For instance, it is not sensible to issue tickets for a service that can be consumed without specific limitations. If, for example, a consumer enters into a contract for an online newspaper subscription without a specific time limit, it is not sensible to issue an admission ticket for each time the newspaper is accessed. In my view, the most sensible use of tickets is to issue them for a single or limited number of access rights, such as downloading a specific resource or streaming a certain video.

As regards the execution of contract rights, a DRM system has the ability to combine electronic tickets with other mechanisms. Figure 4.5 depicts the case in which a customer purchases a subscription for an online newspaper, where electronic tickets are combined with *direct processing* of access rights. As described in previous examples, upon contract conclusion an electronic ticket is derived from the contract for the right to collect money. This ticket is then forwarded by the processing DRM system to the operator of the online newspaper. In contrast to the contract right of the seller, the contract rights of the customer will not be issued as tickets, but directly processed by the the DRM platform. In this scenario the DRM platform is most likely a secured web server that converts the contract rights to access right on the web server. After the access rights are processed the consumer has the ability to access the online newspaper according to the terms and conditions of the contract (e.g. for one month).

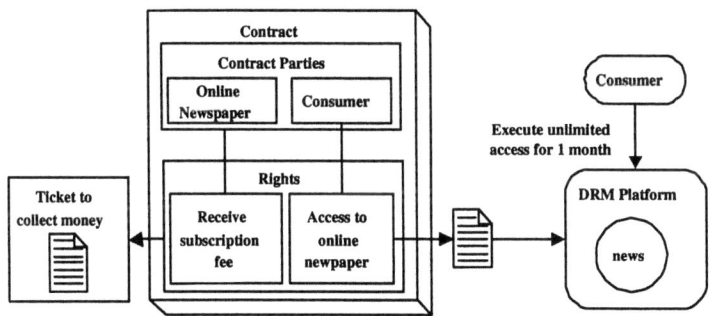

Figure 4.5: Combination of tickets and direct rights processing

Note that Figure 4.5 differs from Figure 4.4 in Section 4.3.1, as it addresses not only the derivation of contract right to tickets, but partly describes their processing. The contract rights in tickets as well as the contract rights that are directly processed on the DRM platform contain rights expressions written in a rights expression language. Section 7.2 in this thesis addresses the implementation of directly processing electronic contracts. In Section 2.3 an application and the processing of electronic tickets has been described. In the sample DRM system the booking results in an electronic contract. After the payment is settled, the *Content Preparation* module adds a license (i.e. an electronic ticket) to the secure container. The license is later processed by the secure viewer on the consumer PC.

4.4 Contract Objects and Contract Use

In general, depending on the content of documents, different general document types can be identified. For example, a *recipe* is a document type and usually comprises the ingredients for a certain product and the work instructions for the production process. Likewise, a handwritten as well as an electronic *contract* is a document type that is characterised by containing one or more parties that exchange rights or products under certain terms and conditions. This section introduces a contract data model, that includes typical contract objects and their interrelations. Here the term *contract objects* refers to instances of classes that occur in contracts.

Contract objects can be subdivided into *core objects*, and additional *scenario-specific objects*. In Section 4.4.1 the interrelated core objects of electronic contracts and their attributes are introduced. Electronic contracts can be applied in various usage scenarios (see Section 4.4.2). Each usage scenario requires a distinctive agreement category. Each agreement category may require extra information in the electronic contracts. Scenario-specific objects, and their are addressed in Section 4.4.3.

4.4.1 Core Contract Objects

This section introduces three abstract contract objects which can be seen as *core objects* of electronic contracts. These three core objects of electronic contracts are: *Party, Resource* and *Permission*. The definition of these abstract core objects was influenced by earlier information models [Ian01], my experience with current rights expression languages which often apply similar approaches [DWW03, Ian02b], and the investigation of projects in which electronic contracts are used (e.g. the COLIS project[4]).

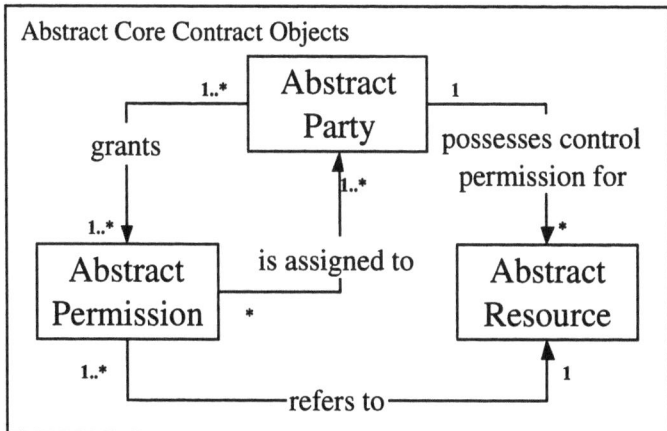

Figure 4.6: The abstract core objects of electronic contracts

The core contract objects are interrelated as follows: Specific parties ("rights holders") possess intellectual property rights for one or more specific resources, such as books, software, music files, or digital videos. A party in possession of such property rights is authorised to grant usage permissions to other persons (customers, or beneficiaries). As a result, permissions are assigned to parties. Each permission refers to one particular or one specific type of resource, and one or more permissions may exist for each resource.

Each of the core contract objects shown in Figure 4.6 comprises a number of attributes. The required object attributes and their relevance are explained in the following:

- *Party* is a mandatory contract object that appears at least twice in each contract. A *Party* instance represents contracting parties, e.g. consumer and seller, and other contract–related persons. With respect to digital contracts, different party types can be distinguished: A *rightsholder* is a party that holds rights on the respective contract resource and may grant those rights to another contract party. A

[4]See: The Collaborative Online Learning and Information Services (COLIS) Project, http://www.colis.mq.edu.au/

consumer is the party that receives rights from the rightsholder. A *beneficiary* can be a third party that may actually execute the rights of the contract on behalf of the consumer. Parties must always be identified by a unique *id*. In general, different types of values can be used to identify a contract party, e.g.:

- A (globally or at least locally) unique identifier that identifies a certain *individual*, e.g. an X.500 distinguished name [IT93a] or a Kerberos [SNS88] established identity.
- A unique identifier that identifies a certain *party type* (e.g. as defined in MARC 21 role code list [MAR03]). A party type represents a number of individuals sharing one or more common characteristics. For example, a party type "faculty member" could represent each faculty member at the Vienna University of Economics and BA.

Note that additional scenario–specific attributes can be assigned to parties, such as *name, role, position, age, credit standing, profession,* etc.

- *Resource* is a mandatory contract object which denotes the objects of the agreement that are the actual digital goods or services. Resources are likewise identified by a mandatory unique *resource id*, as well as by optional metadata. Possible identifiers for digital goods are a uniform resource identifier (URI) [BL94] or a digital object identifier (DOI) [Nat00]. The scenario–specific metadata attributes may supply details on the resource, for example *book title, author, isbn number, description, size, file format, author's remarks*, etc.

- *Permission* is a mandatory contract object which appears at least once in a contract. A *Permission* represents the concrete usage rights granted/assigned to the consumer as a result of the conclusion of the contract. Permissions express usage rights (e.g. play music file, print document, etc.) and may also comprise attributes describing (informal) copyright information or (informal) derivative work rights. A permission always represents at least an $\langle operation, object \rangle$ pair describing an operation that can be invoked on a specific resource (or object), i.e. the two mandatory object attributes of a *Permission* software object are operation and object. A simple example of a permission is: $\langle print, researchpaper \rangle$. Typical operation values include terms such as *print, play, copy, modify*. Permissions stated in a valid contract can be enforced either legally or electronically (by a software service). Currently, there is no ongoing standardisation initiative for operation terms. Therefore, operations can not be uniquely identified, which may lead to ambiguous interpretation (see Section 4.6).

Generally, ids have to be uniquely identified for each type of application, even if no globally unique ids, such as x509, or DOI, are used or available. For example, operation terms that are used in an application can be defined in a rights data dictionary (see Section 3.3.2). Rights expression languages (RELs) (see Chapter 3) that are used for contract representation facilitate the expression of the core contract objects mentioned above (e.g. [Ian02b], [DWW03]).

4.4.2 Sample Usage Scenarios for Electronic Contracts

This section presents an overview of potential usage scenarios for electronic contracts. The selection of usage scenarios is based on experiences in the field of electronic contracts, an additional analysis of the literature on rights expression languages [DWW03, Ian02b], research papers [MSM01, GSSS00] as well as projects dealing with the management of electronic contracts (e.g. the COLIS project[5]).

Access Control

Contracts contain information on permissions concerning digital goods or services. This information is suitable to serve as basis for access control. Theoretically each contract right can be transformed into an access control statement. Primarily, access rights to digital resources that are stated in electronic contract are suitable to be processed in an access control mechanism. The information that is processed in this usage scenario depends on the access control model (e.g. Role–Based Access Control or Discretionary Access Control) that shall be applied.

Accounting

Electronic contracts are the documentation of an agreement between two parties over assets or services. Usually, in an agreement duties respectively payments are stated that have to be settled by the contracting parties.

---

[5]See: The Collaborative Online Learning and Information Services (COLIS) Project, http://www.colis.mq.edu.au/

Therefore, electronic contracts can be an information source for accounting services. For example, every time a contract is concluded, the monetary duties are transferred to open positions in the accounting service.

Intellectual Property Rights (IPRs) Protection

From a legal perspective, content owners market their IPRs to customers. Electronic contracts provide a means for the content owner to specify the extent to which the content may be used. As the IPRs specified in electronic contracts can be (semi–)automatically enforced, the IPRs of content owners are protected. Therefore, on a technical level, this can be seen as a special case of the usage scenario access control.

Customer Relationship Management (CRM)

The overall body of contracts concluded can represent a valuable data pool for marketing activities. Information on purchasing habits of customers, that is which goods or what type of services customers usually demand, can form an information basis for marketing activities such as one–to–one marketing or personalised marketing within the framework of CRM. For example, personalised goods or services can be offered to the respective customers on the basis of their recorded contract history.

Workflow Management

To a certain degree, electronic contracts can be used to specify workflow process information. This process information can be used to control certain task sequences in an information system. For example, let us assume that a contract states that a right is granted after a certain amount has been paid to the content provider. This information reveals a sequencing of tasks that have to be executed and can be used in the workflow process, as follows: an incoming payment event related to an electronic contract initiates the assignment of rights to the respective consumer(s) or contract party.

This section has spanned a relatively broad range of possible uses for electronic contracts. However, it does not claim to be complete and could be extended at reasonable expense. This thesis especially focuses upon the usage of electronic contract in access control services.

4.4.3 Scenario–Specific Contract Objects

The information to be included in electronic contracts varies depending on the scenarios they need to satisfy. For some scenarios not all contract information can be sufficiently represented by the core contract objects or by an extension of their attributes. For example, when using electronic contracts for accounting, the payment conditions and banking details have to be included. Consequently, scenario–specific objects and attributes have to be identified and added to the data model. The additional objects have to be set in relation to the core contract objects. The number of objects and attributes that satisfy one usage scenario is called *agreement category*. Naturally, the agreement category comprises the core contract objects and scenario–specific contract objects. If in one specific application an electronic contracts has to satisfy several usage scenarios, e.g. access control, accounting, and CRM, the number of contract objects and attributes potentially grows with each additional scenario. More precisely, for each additional usage scenario objects have to be added that are not yet covered by earlier agreement categories. Contracts objects that conform to a specific application (that possibly includes several usage scenarios) are denoted as *application–specific* contract objects.

Each application then accesses specific contract information, i.e. contract objects and their attributes, when the contract is processed. Figure 4.7 illustrates various agreement categories (AC1-5) and their overlapping as well as how different software services access electronic contracts in order to fulfill specific usage scenarios. The contract objects include the core objects (white rectangle) and additional, interrelated, scenario–specific objects (colored rectangles). The resulting data model of an electronic contract is called application–specific data model. An application–specific data model is described in more detail in Section 4.6. It could be argued that it is sufficient to simply state the ids of resource, party, permissions, constraints, etc. in an electronic contract and query the scenario–specific attributes from a database. However, this proceeding would contradict the original goals that have been defined for this thesis, e.g. to support contract transparency for contracting parties in e–commerce.

The following section describes how additional objects and attributes can be identified for a number of (additional) usage scenarios.

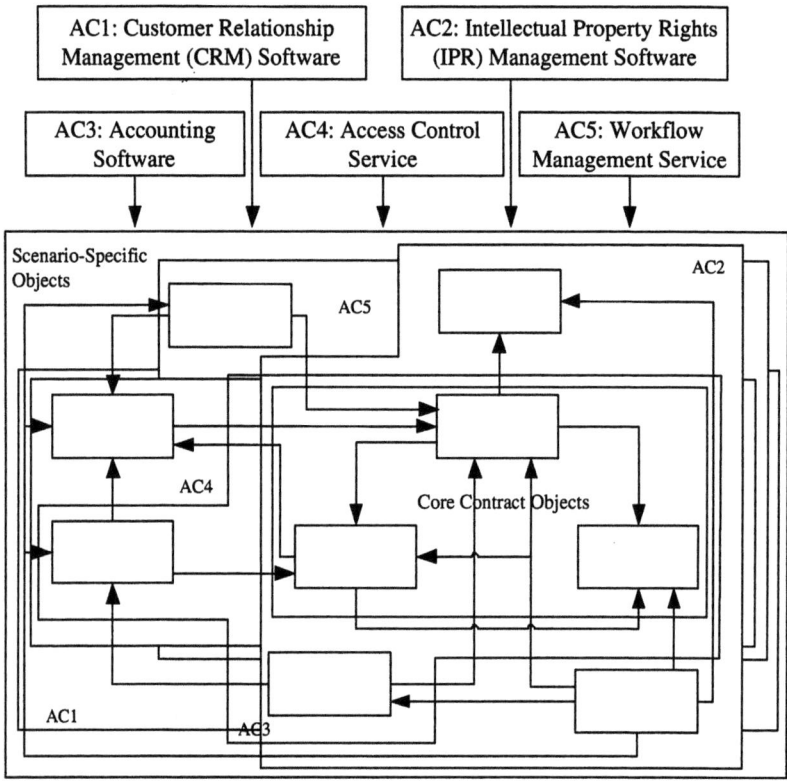

Figure 4.7: Various usage scenarios for electronic contracts

4.5 Contract Modelling and Creation

This section deals with the modelling of electronic contracts and their formulation in rights expression languages tailored to specific usage scenarios. One main reason why the formulation of electronic contracts in rights expression languages and their interpretation make sense is that two or more DRM systems respectively DRM system components use the electronic contracts to exchange rights expressions. Contracts formulated in rights expression languages are suitable to serve as an interface between DRM systems and DRM system components, if their contents meet the specifications of the respective rights expression language (see Chapter 3) that the two systems have agreed on. Nevertheless, for most applications it is not sufficient to simply agree on a REL. If the electronic contract shall be reliably processed in a sensible application, such as access control, the DRM systems additionally need to agree on application policies (see Section 3.3.2) and the contract content. The contract content is defined by contract objects as described in the previous section. This section addresses how application-specific contract objects are identified.

A concrete implementation that processes electronic contracts in a specific usage scenario is referred to as *software service*. Software services are, for example, accounting software, access control mechanism, etc. Section 4.5.1 identifies information which is required to satisfy two different usage scenarios. Section 4.5.2 derives a contract data model from the required information by identifying contract objects, their attributes and relations. In Section 4.5.3, a process for the tailored composition of electronic contracts is introduced. This process supports the composition of contracts, i.e. assembling contract objects and their attributes, tailored to their intended use (i.e. the usage scenarios a contract is to be applied to).

4.5.1 Required Information for Specific Software Services

As mentioned above, all information in electronic contracts and their respective uses should be clearly defined in advance in order to facilitate the automatic processing of contract information. Before identifying the required information of electronic contracts, the intended usage scenarios have to be defined. Each scenario requires a certain number and type of contract objects and attributes in order to process contract data properly, more precisely, *software methods* that accomplish the respective usage scenario in a well–defined sequence require certain attributes. Therefore, the first step towards tailored contracts is to identify the information that the software methods of each scenario require. In the following requirement analysis, it is assumed that the respective contract shall be processed in an access control mechanism and in an accounting software. For each required attribute it has to be defined whether it is a mandatory or an optional attribute for this application, how often the attributes will occur in the contract, and to which other attributes it is related.

Required Information for Role–Base Access Control

Access control mechanisms aim at regulating the access of users (subjects) to resources. When using the discretionary access control (DAC) approach, access permissions are directly assigned to the users. For example, the permission *read book #"The future of ideas"* is directly assigned to subject *sguth*. The role–based access control (RBAC) approach assigns usage permissions to *roles* rather than to subjects [FSG+01, SCFY96]. The roles are then assigned to subjects. Thus, users receive permissions transitively via their assigned roles (see Figure 4.8). For example, the permission *read book #"The future of ideas"* is assigned to the role *researcher*, which in return is assigned to user *sguth*. Roles can be arranged in role hierarchies, in which more powerful roles (senior roles) inherit permissions (and constraints) from subordinate roles (junior roles). A role hierarchy is a directed acyclic graph. Roles are a convenient means to assign and manage permissions.

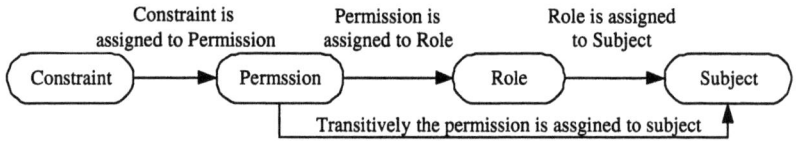

Figure 4.8: Assigning permissions in RBAC

Permissions can also be associated with *constraints* [NS03b]. One specific type of constraint is a *precondition*, i.e. a premise which has to be fulfilled before a permission can be granted/assigned to a specific subject. Other types of constraints might restrict permissions, for example to a certain time interval or to a specific user or device. For instance, prepayment might be necessary in order to receive the permission to play an audio file (assignment constraint), and once the fee is paid, this particular permission might be exercised a limited number of times (authorisation constraint). Thus, to execute the usage scenario "role–based access control" the following methods are used:

- `createSubject(SubjectID)`: Creates a new subject, i.e. the party that receives access rights from the contract.

- `createPermission(Operation ResourceID)`: Creates a new permission from an operation–object pair (see Section 3.3.1).

- `createRole(RoleID)`: Creates a new role.

- `createConstraint(Name Operator Value)`: Creates a new constraint.

- `rolePermAssign(RoleID PermissionID)`: Assigns the created permission to the role.

- `subectRoleAssign(SubjectID RoleID)`: Assigns the created role to the subject.

- `relatedConstraintToPerm(Constraint Permission)`: Assigns the created constraint to the permission.

These methods have to be provided with contract information to implement the access rights from an electronic contract to the role–based access control mechanism. In other words, the parameters of the methods are the required information for the usage scenario role–based access control. In the following the RBAC–specific contract information is described in detail:

- *Subject Type.* To identify which of the contract parties receives access rights, each contract party must be further specified by a type, e.g. customer, seller.

- *Subject-ID.* For all contract parties that receive access rights to resources via the contract, a user id is required (e.g. sguth, mstrem). A consumer subject id occurs at least once in a contract and is mandatory data that uniquely identifies the users in the system to which the access control assigned shall be assigned.

- *Resource-ID.* For all resources that access shall be granted to, a resource id is required (e.g. music–file#12345). Resource id occurs at least once in a contract and is mandatory data that uniquely identifies the resource in the system where the resource is stored and secured by the respective access control mechanism. The resource id can also identify a set of resources, such as a folder or a certain type of resource.

- *Operation.* The operation is mandatory data for RBAC that has to occur at least once in a contract. Remember that in the access control community a permission is a pair consisting of a certain operation (e.g. play) and an object (or resource, e.g. music–file#12345). If more than one permission is defined and the contract comprises several contract parties, the relations between parties and permissions have to be stated unambiguously .

- *Role.* The role is mandatory data in RBAC, unless the access control mechanism can also handle discretionary access control (see Section 4.4.2). A role that is "known" by the respective RBAC service has to be assigned to each party that shall receive access rights. A role is a named collection of users and permissions, and possibly other roles [San96]. Role names sometimes resemble user group names, such as *student, employee,* etc.

- *Constraint.* A constraint specifies that certain context attributes must meet certain conditions in order to grant a specific permission. For example, a constraint may specify a date until which the permission is valid. A constraint can be assigned to no, one, or several permissions and has at least three attributes, two operands and one operator, for example, *name, operator,* and *value,* e.g. $\langle date, <, 12/31/2004 \rangle$. Constraints are optional data.

Required Information for Accounting Services

From Section 4.4.2 it can be learned that, for example, every time a contract is concluded, the monetary duties can be transferred to an open position in the accounting system. To implement the use of electronic contracts in accounting services the following method has to be called:

- `createOpenPosition(PartyID ResourceID Duty Conditions)`: Creates a new open position in the accounting sytem.

This means that the attributes Party–ID, Resource–ID, Duty, and Term and Condition are required:

- *Party ID.* For all parties that are involved in monetary transactions an id has to be specified (e.g. sguth, mstrem). Party id is mandatory data that uniquely identifies the user in the accounting system and occurs at least once. In accounting software users are related to payment obligations (duties) and to the resource they have purchased.

- *Party Type.* Each contract party must be further specified by a type with respect to payment relations, e.g. customer or seller.

- *Resource ID.* Resource id is optional data that, if available, uniquely identifies the traded resource in the accounting system (e.g. music-file#12345). Although a duty can be booked without the related resource or service id, it is reasonable to specify this id, to further specify the business transaction. The resource is related to one or more permissions.

- *Duty.* The duties specify the monetary or non–monetary liabilities between the contract parties or third persons that result from a contract. Duties are mandatory data for the accounting service and for defining the type (e.g. amount of money), value (e.g. 1000,00), and attribute (e.g. €) of the duty. The goods might also be bartered or paid for with artificial credits, which are alternative occurrences of a monetary *duty*. However, one or more duties can be specified in a contract. Duties are related to a certain contract party and can be related to constraints.

- *Terms and Conditions.* A duty is often afflicted with terms and conditions, such as *"the payment has to be settled until 31st December 2003."* Such terms and conditions are optional data and can also be expressed as triples, like the constraints in the role–based access control example, e.g. $\langle settlement\ datetime, <, 01/01/2004 \rangle$. None, one, or many conditions can be assigned to a duty and one condition can be assigned to several different duties.

4.5.2 Modelling Scenario-Specific Contracts

In Section 4.4 the abstract core contract components *Abstract Party*, *Abstract Permission*, and *Abstract Resource* have been introduced. Section 4.4.3 explains why for each agreement category it is necessary to extend the core objects with scenario–specific objects to ensure the sophisticated processing of electronic contracts in these scenarios. In this section these scenario–specific objects will be identified on the basis of the attribute analysis in Section 4.5.1.

The required contract objects for the agreement category "role–based access control", are *Party, Resource, Permission, Constraint,* and *Role*. The attributes *resource-id, subject-id, subject–type* and *permission* can be represented with the contract objects *Resource, Party,* and *Permission*. Thus, for access control purposes the core objects have to be extended by the objects *Role* and *Constraint*:

- *Role* is a mandatory access control-specific contract object. Its single attribute *name* is storing the role name. Roles are related to *Party* objects.

- *Constraint* is an optional, scenario–specific contract object. This object type provides the three attributes *type, operator,* and *value.* Constraints are related to *Permission* objects.

With these three new contract objects the agreement category for role–based access control service can be represented. The same procedure has to be accomplished for the scenario–specific attributes of the accounting software. The attributes *party id, party type,* and *resource id* can be represented by the *Party* respectively the *Resource* object. As the object attribute *terms and conditions* can be expressed with the same attributes as the RBAC constraints, no additional objects or attributes are required for this attribute. Finally, to satisfy the accounting service a new contract object of the type *Duty* has to be added.

- *Duty.* The *Duty* object is comprising the attributes *name, value,* and *attribute. Duties* are related to *Party* objects, and can be associated with *Constraint* objects.

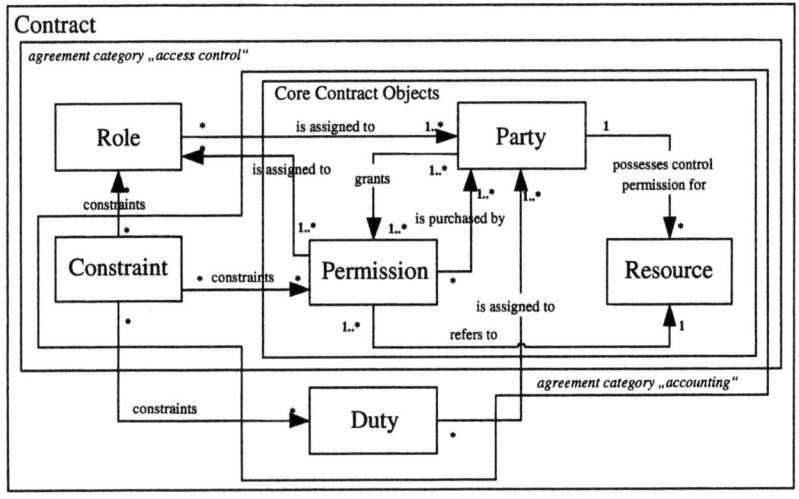

Figure 4.9: Application–specific data model

In a contract that has additional scenario–specific contract objects, such as *Role, Constraint, Duty,* and *Contract,* the relations between the contract objects change, respectively new relations have to be added. A contract that is processed by an access control mechanism or and accounting software and automatically results in access rights to protected resources respectively open positions must be secured, i.e. it is necessary to check if the contract is valid, e.g. with an electronic signature. Such external characteristics of the contract do not belong directly to the respective usage scenario and are implementation–dependent. The additional contract object *Contract* is comprising such extra attributes for eventual validity checking, i.e. it can store digital signature, physical location of the contract conclusion, date, etc. The *Contract* object also aggregates the remaining contract objects and therewith provides a means to assign the contract objects to one specific contract. Figure 4.9 depicts the application–specific data model, i.e. the resulting contract objects and their relations (also called contract schema) that are required for the usage scenarios *RBAC* and *accounting*.

In the example, the *Constraint* object is related to the *Permission* object as well as to the new objects *Role* and *Duty*. Constraints are capable of narrowing *Permissions, Duties* and *Roles*. A *Constraint* can be assigned to an infinite number of *Permissions, Duties,* and *Roles* and vice versa. The *Role* objects are now assigned to *Party* objects. The "is-assigned-to"–relation, depicted in Figure 4.6 between *Permission* and *Party* is no longer required for the current usage. *Permissions* are now assigned to *Role* objects. A *Duty* can be related to one or more *Parties*, and one *Party* can be related to one or more *Duties*. A new relation between *Party* and *Permission* indicates which *Permissions* have been purchased by *Parties*.

With these contract objects the agreement categories for both usage scenarios can be represented. Figure 4.10 shows instances of the object types, their attributes and actual values. The values have been taken from the examples of each attribute from the previous section. Furthermore, Figure 4.10 shows the mapping of the respective instances and their attributes to the corresponding software services. Each software service may require several attributes from different instances; contract objects (and attributes) might be used in one or more software services. For example, *role* is solely processed in the RBAC service (or the seller information in the accounting software), and some attributes are processed in both, e.g. the customer id and the permission. Due to clarity reasons the aggregation function of the contract object is not illustrated in Figure 4.10. Please note, that the illustrated case is an example, and other access control and accounting software might use different attributes.

With a growing number of software services, the required objects and attributes for the contract will increase as well. Figure 4.7 in Section 4.4.3 can be seen as an extended example where various usage scenarios are mapped

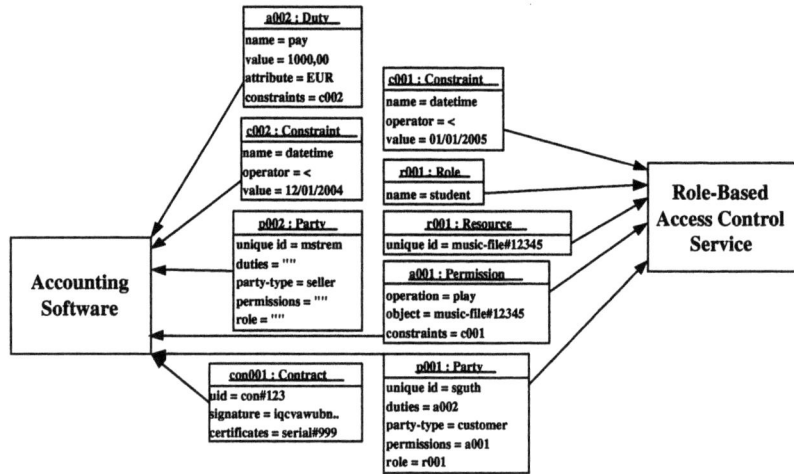

Figure 4.10: Example of mapping of objects instances and their attributes to software services

to a number of agreement categories respectively contract objects. The modelling of electronic contracts is important for communication and implementation purposes (see Section 4.6). The next section introduces a process that aims at ensuring that electronic contracts include all required objects identified in Section 4.5.1.

4.5.3 Scenario–Specific Contract Composition

This section describes a conceptual framework which enables the composition of electronic contracts formulated in any rights expression language and tailored to the requirements of specific usage scenarios. In order to include application–specific objects and attributes in as contract document, it is necessary to know the *usage scenarios* for this particular contract in advance, i.e. in order to ensure that electronic contracts contain sufficient information to satisfy the requirements of specific usage scenarios, the contracts need to be tailored with regard to their intended use(s). Subsequently, a simple process for the tailored composition of electronic contracts is introduced that adds contract attributes to a contract document based on its usage scenarios respectively its agreement categories. Figure 4.11 depicts such a process; the respective activities are described below.

- *Identify relevant usage scenarios*: In this activity, the list of usage scenarios (see Section 4.4.2) is identified to specify the intended use of the contract under consideration. In other words, the estimated use of this particular contract is defined, for example access control and accounting.

- *Identify software methods*: Here, the software methods of each scenario that finally process the contract data are identified. These software methods, executed in a well–defined sequence accomplish a certain usage scenario.

- *Identify required attributes*: The parameters of the identified software methods are the required attributes of the contract. Therefore, the parameters of each identified software method have to be determined for each usage scenario.

- *Develop agreement category*: For each usage scenario the required contract objects, their attributes and interrelations have to be identified. The resulting data model is the agreement category of the respective usage scenario.

- *Identify application–specific objects*: As shown in Figure 4.9, the agreement categories of the various usage scenarios are overlapping. By combining the various agreement categories, the application–specific contract objects and their attributes are identified. The usage of attributes in software various services is illustrated in Figure 4.10.

- *Append attributes to contract template*: Based on the application-specific objects and attributes a contract template is generated. For this step a tool called *rights expression generator* is used (see Sections 5.1.2 and 6.2). In the analysis shown in Section 4.5.1 the mandatory and optional attributes and their occurrences have been identified. The characteristic whether an attribute is mandatory or optional has to be taken over for the creation of the contract template. The contract template is formulated in the preferred rights expression language. When creating the template, it must be considered that the contract information can be unambiguously mapped to the contract data model, as defined earlier. This is a prerequisite for the reliable processing of the contract (see Section 4.6).

- *Fill in contract*: In this activity the different attributes with actual values are filled in, i.e. the party ids, the party types, the resource

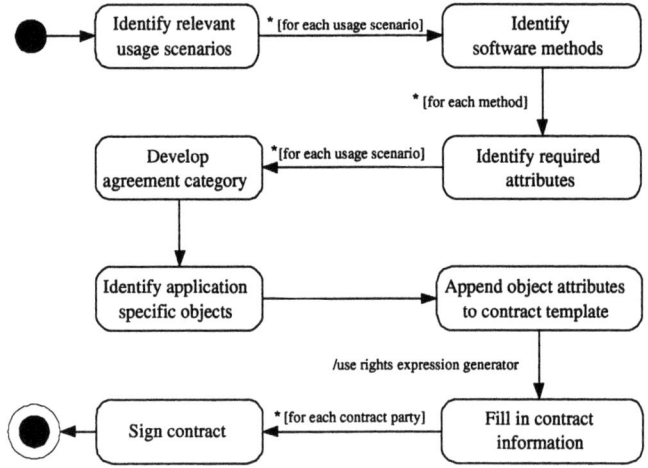

Figure 4.11: Composing tailored electronic contracts

ids, the roles, the permissions and duties, and the terms and conditions are included in the contract. Note that the contract parties should have the ability to add additional contract information that has not been identified as mandatory. The contract template ensures the availability of, but shall not restrict the contract to the mandatory attributes. At this point the contract is still in the negotiation or offer phase (see Section 4.1). All contracting parties have to agree on the actual attribute and their values before the contract can be signed. If necessary, the contract is modified until an agreement is reached (see also Figure 4.2).

- *Sign contract*: In the final step, the contract has to be signed by each contracting party in order to be considered as valid. At this point, the contract reaches the 'Conclusion' stage in its life cycle. Without the signatures of the contract parties the contract is not valid (see Section 4.8) and will not be processed.

A framework that supports the above mentioned process, i.e. which generates the contract template and transforms the filled out contract to a REL instance, can be built on top of the rights expression exchange framework designed in Chapter 5.

The tailoring process increases a contract's enforceability [GK02]. For example, a contract consisting of the contract objects shown in Figure 4.9 has the potential to be fully electronically enforceable in terms of access control services and accounting. However, if a contract is not designed as proposed above, it can still be a valuable source of information for specific applications, such as sales statistics and customer relationship management. In such cases, electronic contracts can be processed in 'unstructured' ways, such as in data mining procedures. Then, usage scenarios will be less likely be provided with all required information. Accordingly, the enforceability of such contracts is low.

Note that the question of who decides on the contract usage and consequently on the contract content is a serious privacy matter. This privacy topic also requires a discussion of organisational, management, and privacy issues in contract composition, because the contract content has to be agreed upon by all contracting parties. For example, on the one hand consumers will demand that their personal information is handled confidentially, while on the other hand the marketing department will be interested in personal information for CRM purposes. This issue occurs in every step of the supply chain, in which electronic contracts are applied.

4.6 The Generic Contract Schema

This section deals with the processing of electronic contracts. It introduces a further development of contract models from the previous section and introduces a representation of contracts in a generic data model.

4.6.1 Definition of Terms

In the previous section the three core elements have been introduced. The interrelated objects *Party*, *Permission* and *Resource* are the minimum constituents of every concluded contract, i.e. they build the core data model of electronic contracts. To these core objects, additional objects can be added to represent application–specific information required by the usage scenarios of the contract.

> The *generic contract schema* (generic CoSa) is the theoretical idea of a contract schema that is capable of representing the to-

tal of (electronic) contracts. The entirety of usage scenarios, and thus all scenario–specific extensions of the core elements underly such a generic contract schema. Consequently *any* (electronic) contract can be mapped to the generic CoSa. The generic CoSa is thus an abstraction layer of various representations of electronic contracts, such as ODRL, XrML, other rights languages, or even contract runtime models. The generic contract schema shall provide for a higher level of standardisation and openness in DRM systems that process rights expressions. To access contract data, the generic CoSa can be queried via the CoSa *interface*. The CoSa interface is a generic application programming interface (API) which is independent of usage scenarios, allows to query all contract data, and thus facilitates a standardised processing of contract data in software services.

The implementation of the generic contract schema is a considerable challenge and probably technically impossible. It would require knowledge about all today's and tomorrow's contract usage scenarios as well as the accordingly needed contract objects and their relations. Yet, the generic CoSa interface is practicable and will be presented later in this Section. However, with some restrictions the generic contract schema can be implemented and helps facilitating and standardising the processing of contract data:

- **Domain–specific CoSa.** The domain–specific CoSa covers a great number of deal of popular contract objects and their relations in a specific domain (e.g. education, music industry). The domain–specific contract schema would permanently undergo further development (such as the the Learning Object Metadata (LOM) standard [IEE02]) and an independent organisation (such as the Learning Technology Standards Committee (LTSC)[6]) would watch and control the development of the contract schema. The domain–specific CoSa would also provide guidelines for the contract object attributes and their allowed values. For example, the permitted attributes of *Resource* objects are all defined in the LOM standard, or the permitted attributes for *Party* are attributes defined in the vCard [HF98] standard. If not already defined in the metadata standard, the permitted values of attributes need to be stated as well, e.g. the attribute *Identifier* may comprise ISSN [ISO98], ISBN [ISO92], and DOI [Nat00] compliant identifiers. The extension of the contract schema by new objects

---

[6]See: http://ltsc.ieee.org/

	Domain–specific	Application–specific
Covered Scenarios	prevalent ones	all
Reliability	< 100%	100 %
Flexibility	good	poor

Table 4.1: Characteristics of application–specific and domain–specific CoSa

or attributes has to be requested and publicly discussed. After this process, new objects can be added to the schema by relating them to existing objects. Technically, this means that it can not be guaranteed that contract data which is not covered by the domain–specific CoSa can be reliably processed respectively enforced. Still, this approach offers high flexibility for actors in the educational domain, as electronic contracts can be easily processed in additional usage scenarios (see Table 4.1). Also by implementing the domain–specific CoSa, actors in that domain can ad hoc use and provide services, such as trading, booking, and rendering of electronic goods.

- **Application–specific CoSa.** The generic contract schema, as defined above, also holds for a specific, closed application, i.e. an application specific data model (such as shown in Figure 4.9). One application can comprise several usage scenarios, e.g. access control and accounting. The application–specific CoSa is fixed in advance; apart from the contract objects this also includes the object attributes and their permitted values. As in the domain–specific CoSa already existing description standards, such as LOM, Dublin Core or vCard, can be reused for this purpose. All contract data can be reliably processed in the designated usage scenarios. The disadvantage of this approach is caused by its prerequisites (predefined and fixed data model) that lead to a poor flexibility. The processing of electronic contracts in additional usage scenarios or extended object attributes requires changes in the application–specific contract schema which, in return, requires software modification of the contract interpreter.

The characteristics of the domain–specific and the application–specific CoSa are opposed in Table 4.1.

4.6.2 Application–Specific CoSa Example

In the following a simple example of an application–specific CoSa is presented, in order to address further technical details of the CoSa approach.

In this sample application it is assumed that the processing in intellectual property rights (IPR) management services is the only usage scenario of electronic contracts. This usage scenario requires additional information about the contract itself, i.e. it is necessary to store the physical location where the contract is concluded to determine the legal venue. Therefore, the core objects have to be extended by the object *Contract*. The *Contract* object aggregates all other objects and therefore has to be related to all core objects (*Party, Permission,* and *Resource*). Figure 4.13 illustrates the objects and their relations of the agreement category for IPR. Because the application only includes one usage scenario, the agreement category for IPR is concurrent with the application–specific CoSa.

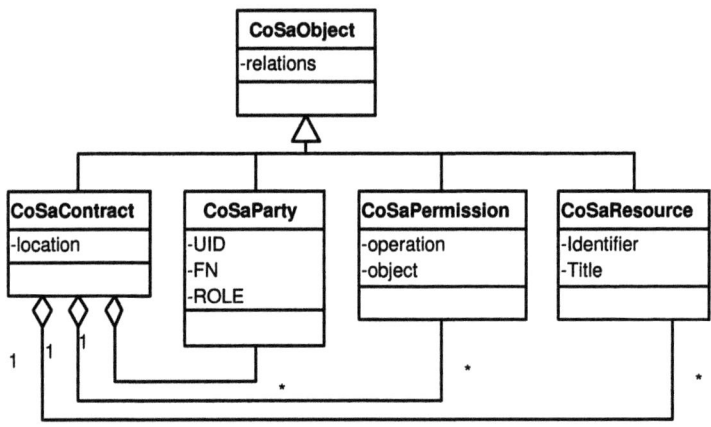

Figure 4.12: Class diagram of an application–specific Contract Schema

The resulting class diagram of the application–specific contract schema is depicted in Figure 4.12. This figure also reflects the functionality of the *Contract* object to aggregate all other objects of one particular contract. The class *CoSaObject* comprises common instance variables and methods of all contract elements (here, the attribute `relations`).

Figure 4.14 shows instances from the application–specific CoSa classes. In fact, it depicts all instances that are part of a small sample contract (see ODRL serialisation in the listing below) and that have been mapped to the application–specific CoSa.

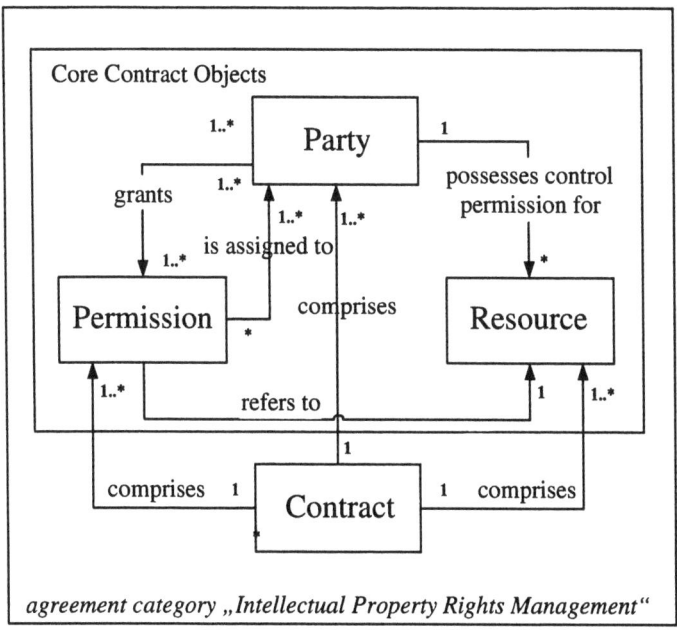

Figure 4.13: Application–specific contract schema

```
<rights>
  <agreement>
      <context>
          <pLocation> Germany </pLocation>
      </context>
      <party>
         <context>
           <uid> #sgraf </uid>
           <name> Steffi Graf </name>
         </context>
         <rightsholder/>
      </party>
      <party>
         <context>
           <uid> #bbecker </uid>
           <name> Boris Becker </name>
         </context>
      </party>
      <asset>
```

120

```
            <context>
                <uid> ebook#123 </uid>
                <name> International Tennis Rules </name>
            </context>
        </asset>
        <permission>
            <print/>
        </permission>
    </agreement>
</rights>
```

The contract has two parties, the consumer Boris Becker with the (system wide) unique id #bbecker and a seller Steffi Graf with the unique id #sgraf. Steffi is selling a print permission for an Ebook on international Tennis rules to Boris. The Ebook has the unique id ebook#123. The contract itself was concluded in Germany. Classes from the application–specific CoSa have two different kinds of attributes: *contract attributes* and *intrinsic attributes*:

- *Contract attributes.* In contract attributes the factual contract information is stored. Here, *FN, UID, ROLE* are contract attributes for the object *Party* and *Identifier and Title* etc. are typical contract attributes for the object *Resource*. The application–specific contract schema is reusing existent, standardised vocabulary for the contract attributes. The *Resource* attributes are expressed by the Dublin Core [Dub01] vocabulary , and *Party* attributes are taken from the vCard

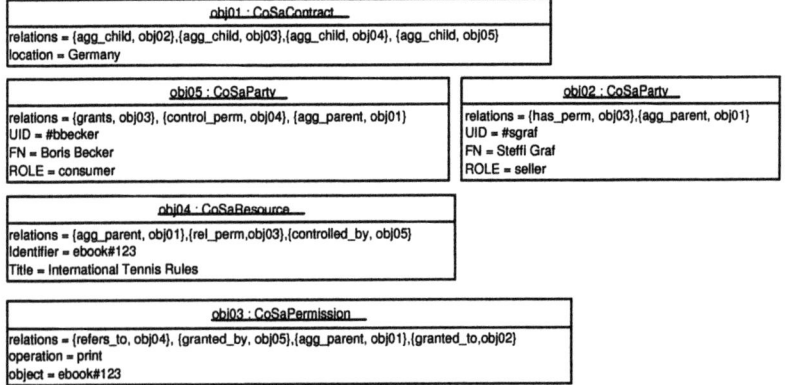

Figure 4.14: Instance of an application–specific Contract Schema

From	To	Relation	Role (from)	Role (to)
Permission	Resource	refers to	refers_to	rel_perm
Contract	Resource	comprises	agg_child	agg_parent
Contract	Party	comprises	agg_child	agg_parent
Contract	Permission	comprises	agg_child	agg_parent
Party	Resource	contr. permission	control_perm	controlled_by
Party	Permission	is assigned to	has_perm	granted_to
Party	Permission	grants	grants	granted_by

Table 4.2: Possible role names in application–specific CoSa

standard [HF98]. We are not aware of initiatives that define usage and access rights or general contract attributes customary in a line of trade. The contract attributes are extensible by additional vocabulary sets.

- *Intrinsic attributes.* Intrinsic attributes are necessary to express the relations between contract, objects as defined in the contract data model. The contract schema defines the attribute **relations** to store all relations that the actual objects has to other objects in the contract schema. This attribute comprises a list of ⟨*type, CoSa-object*⟩-pairs. Each pair represents one relation, for example, the relation ⟨*refers_to, obj04*⟩ of obj03 expresses that the permission obj03 is granted to the resource that is represented by object obj04. All possible relations for this specific application can be derived from the data model of the contract schema (see Figure 4.13). One relation in the data model results in two relation roles (names for both ends of the relation), the role names have to be unique. For example, the relation with the label **refers to** between the CoSa objects *Resource* and *Permission* in the data model results in the two roles **refers_to** and **rel_perm** (related permission). A list of all roles from the current small example can be found in Table 4.2.

4.6.3 The CoSa API

The generic CoSa aims at supporting the standardised processing of electronic contracts. Therefore, in the previous section an application–specific contract schema has been developed and described. To also access the contract schema in a standardised way, an application programming interface

(API) is required that allows to query every information from the contract schema and process it in subsequent software services. In the following, such application programming interface, the CoSa API, is introduced, i.e. the CoSa API. The methods only are presented with a short description, for a complete reference of the CoSa API please refer to Appendix B (Chapter 10.1). The method prefix `cosa` represents an instance of a class that is implementing the CoSa API. The `cosa` instance is initiated with an electronic contract and holds all contract information in runtime mapped to the application–specific CoSa. The question marks tag parameters that are optional. The method calls are coded in XOTcl [NZ00b].

- **cosa getObjects object-type**
  This method returns instances of the class *object-type* that exist in the current runtime CoSa. For example, the following command would return the instances `obj05` and `obj02` as a list.

  `cosa getObjects CoSaParty`

- **cosa getRelatedObjects cosaObject ?relation? ?object-type?**
  This method returns all instances related to `cosaObject`. The related instances can be optionally filtered by either the relation, or by the object-type of the related instance, or both. For example, the following command would return the instance `obj03`.

  `cosa getRelatedObjects obj02 has_Perm`

- **cosa getRelObjectTypes cosaObject ?relation?**
  This method returns the object–types (classes) of all CoSa objects related to `cosaObject` optionally filtered for a given `relation`. For example, the following command would return all the object–type *CoSaPermission* of `obj03` which is related to `obj02` in the has_perm–manner).

  `cosa getRelObjectTypes obj02 has_perm`

- **cosa getRelations cosaObject ?relation?**
  This method returns the `relations` attribute content of `cosaObject`.

The content is a list of ⟨relation − typecosaObject⟩-pairs. If the argument **relation** is specified, all pairs are returned of which **relation** equals a certain relation type. For example, the command below would return a list with one pair, namely ⟨agg_parent obj01⟩.

`cosa getRelations obj04 agg_parent`

- **cosa hasRelation cosaObject relation**
  This method determines whether `cosaObject` has a relation to a specific other instance or not. For example, the following command would return **true**.

  `cosa hasRelation obj04 agg_parent`

- **cosa getAllAttributes cosaObject**
  This method returns all attributes of `cosaObject`. For example, the following command would return the list {relations UID FN ROLE}.

  `cosa getAllAttributes obj02`

- **cosa getAttributeValue cosaObject attribute**
  This method returns the value(s) of the **attribute** of `cosaObject`. For example, the following command would return the value **consumer**.

  `cosa getAttributeValue obj02 ROLE`

- **cosa selectObjects list attribute ?value?**
  With this method a list of cosa instances can be filtered with respect to a certain attribute, respectively its value. For example, the following command would return the instance obj05.

  `cosa selectObjects {obj02 obj05} ROLE seller`

This API is short but generic. When new contract objects or relations are added to the contract schema, the API does not have to be changed. You may have noted that the API allows querying contracts but does not offer methods to create or modify electronic contracts. This is due to the characteristics of contracts. Once contracts are concluded, they should not be changed. The concept of the contract schema therefore is not addressing the offer creation, negotiation or contract conclusion phase, but is intended to support the fulfillment respectively the automated processing of electronic contracts.

4.6.4 CoSa Serialisation

The contract schema is the heart of a contract processing framework. As mentioned above, it determines all information of electronic contracts that can be processed in software services. Let us assume that new partners or platforms desire to participate at a domain or application that has defined a CoSa, for example accounting software providers. They would need to know what the contract schema looks like and if it meets the requirements for processing electronic contracts in their accounting software. Additionally, programmers need the names of CoSa objects and their attributes in order to call the CoSa API methods. In short, a generally accepted representation of the CoSa is required to communicate its shape. The resource description framework (RDF) [LS99] is suited for this purpose. The following listing illustrates the RDF serialisation in XML of the application-specific CoSa exemplified in Section 4.6.2:

```
<?xml version="1.0" ?>
<RDF xmlns:RDF ="http://w3.org/TR/1999/PR-rdf-syntax-19990105#"
     xmlns = "http://www.guth.it/CoSa#" >
<RDF:Description about = "Contract Schema for IPR Applications" >
<CoSaObjects name='simple contract type'>
   <CoSaParty>
      <oid/>
      <relations>
            <has_perm/>
            <grants/>
            <control_perm/>
            <agg_parent/>
      </relations>
      <FN/>
      <UID/>
      <ROLE/>
   </CoSaParty>

   <CoSaResource>
      <oid/>
      <relations>
            <rel_perm/>
```

```
            <controlled_by/>
            <agg_parent/>
       </relations>
       <Identifier/>
       <Title/>
   </CoSaResource>

   <CoSaPermission>
       <oid/>
       <relations>
            <agg_parent/>
            <granted_by/>
            <granted_to/>
            <refers_to/>
       </relations>
       <operation/>
       <object/>
   </CoSaPermission>

   <CoSaContract>
       <oid/>
       <relations>
            <agg_child/>
       </relations>
       <location/>
       <uid/>
       <comment/>
   </CoSaContract>
</CoSaObjects>
</RDF:Description>
</RDF>
```

The RDF instance describes the objects and their attributes of the application–specific CoSa. The description contains all information that has been determined in the earlier data model of the specific application (see Figure 4.13). Both the data model and the RDF description serve as communication basis for engineers and developers of applications that process electronic contracts. The following numeration summarises the characteristics and advantages of the generic contract schema concept. The generic CoSa ...

- ... is a concept that enhances openness and interoperability in DRM systems.

- ... aims at standardising the processing of electronic contracts.

- ... serves as abstraction layer for various representations (e.g. rights expression languages) of contract information. The abstraction is achieved by mapping the contract attributes from proprietary representations to generic objects and their attributes[7].

... defines a generic data model for contracts.

... has to be restricted either by domain or application, to be applicable.

... and its API provide standardised access to contract information that has been transformed into the application– or domain–specific contract schema.

Another application–specific contract schema, namely for the usage scenarios access control and accounting, has been developed in Section 4.5.2. An implementation of this contract schema and its processing is shown in Chapter 6.

4.7 Enforceability of Electronic Contracts

Electronic rights enforcement aims at verifying specified usage rights in digital contracts and ensuring their observation both by electronic means. Thus, electronic rights enforcement addresses the enforcement of access rights to electronic goods but not to physical goods. However, the rights to physical goods can still be enforced by legal action. Not all usage rights that can be expressed in digital rights languages can be electronically enforced. For business partners and, above all, for rights holders it is important to know to what extent electronic contracts are enforceable. Identifying and ensuring enforceability of electronic contracts should generate trust as a basis for successful electronic commerce. Therefore, I suggest to divide usage rights into two categories, based on their *enforceability* in electronic contracts:

- *Non-enforceable rights:* These parts of electronic contracts that specify usage rights for resources, which cannot be observed by computer technology.

- *Enforceable rights:* These parts of electronic contracts that specify usage rights for resources, which can be enforced by computer technology. This category must be further specified into:

---

[7]For example, a resource in ODRL is expressed by an *asset*-tag. XrML provides a tag called *digitalWork*. To implement the generic CoSa the two proprietary terms have to be mapped to a generic term, here *resource*

- *Potentially enforceable rights:* these parts of electronic contracts that specify usage rights for resources that are currently not enforceable but have a high potential to be enforced under certain circumstances.
- *Reliably enforceable rights:* these parts of electronic contracts that specify usage rights for resources, which can be reliably enforced as intended by contract parties with existing computer technology.

In order to identify the different levels of enforceability of a given set of rights, clear criteria are needed. Accordingly, the three criteria for electronic contracts to be enforceable are: *availability of required information, availability of appropriate technology,* and *implementation of a trusted environment.*

Availability of required information in the system

The first criterion for the enforceability of an electronic contract is that all required information can be recorded, and/or is available to the system.

Example: As the extensibility of digital rights languages and therefore the range of expressions is limitless, there are no boundaries to information in electronic contracts. The following right expression might not be a common one, but it represents a possible clause in an electronic contract. "The consumer may have access to my entire resources, after s/he has invited my department for a discussion round." The precondition "after s/he has invited my department for a discussion round" is not enforceable, since the necessary information (receive invitation) to check the precondition cannot be recorded by the system.

Control over the usage of resources

A second criterion is the availability of an appropriate technology that permits controlling the specified usage rights for the resource format concerned.

Example: A specified usage right for a resource could be (in words): "The consumer may show the digital video to a class once per semester." We now have all relevant information required to enforce this rights expression. However, in order to prevent the video from being shown more than once per semester, a reliable enforcement technology is needed that is capable of monitoring the usage rights of video formats.

The music industry is currently promoting the development of such technology.

Availability of a trusted environment

The third criterion is the availability of a trusted, i.e. tamper–resistant environment. The term "trusted" here refers to the point of view of the rightsholder who anticipates a license conforming access to resources of the consumer. Rights enforcement in a "trusted" environment is expected to be reliable, i.e. in an non–trusted environment electronic tickets or contracts can be modified, copied, or forged and are then longer enforced as intended.

Enforceabilitiy / Condition	Enforceable		Non-Enforceable
	Reliably Enforceable	Potentially Enforceable	
Information = available Technology = available Environment = trusted	X		
Information = available Technology = not available Environment = trusted		X	
All other cases			X

Figure 4.15: The enforceability matrix

Example: Rights enforcement is easier to implement if the resources are not delivered to the consumer but remain with the provider. The consumer then receives "access rights" to the resource but physically the goods remain stored on the providers' server. For example, when consumers purchase access to an online newspaper, every time they desire to access the latest news, they have to authenticate themselves to the online newspaper. In this case, the system administrator of the platform retains responsibility for access control of the resources. Conversely, rights enforcement is hard to implement if the execution of usage rights is managed by software on the client PC, because then the management of access rights is not the responsibility of the delivery system. We classify the environment of the provider as rather "trusted", and that of the consumer as rather "non-trusted", because no 100-per cent "trusted environment" exists [Fed02]. Due to the characteristics of certain goods it is difficult to provide a trusted, tamper resistant system. For instance, as far as the distributing of music, video, etc. is concerned there is a great amount of recording technology that facilitates making unprotected copies of the resource.

The relation between these three criteria and the three levels of enforceability are represented in matrix form (see Figure 4.15). If all three criteria are fulfilled the electronic contract is enforceable. The process for the composition of tailored electronic contracts (see Section 4.5) respectively the application-specific contract schema in Section 4.6 are means to ensure the availability of all required contract information and facilitate electronic rights enforcement. The DRM sample system (see Section 2.3), for example, addresses the criteria *available technology* (e.g. secure viewers) as well as the *trusted environment* (e.g. secure containers).

4.8 Contract Management Issues

When aiming at processing electronic contracts , a number of technical challenges arise, such as representation of contracts in a machine readable format (addressed in Chapter 3) or defining software objects that are able to store contract information in a software program (addressed in Section 4.4.3). Besides the purely processing issues some managerial questions appear, such as "When is a contract valid?" or "How can I be sure that I bought the content that I intended to buy?" This section aims at drawing attention to these management issues and addressing some relevant ones of them in detail. For the overall focus of this work the identified issues are mainly derived from the access control environment. Figure 4.16 shows the occurrence of some management issues that typically occur along the four phases of the contract life cycle (see Section 4.1).[8] All issues will be addressed in the subsequent paragraphs in detail.

Managing Offer Placement

In the offer placement phase important management issues are *content administration, checking the offer validity*, and *checking the authorisation of the seller*.

---
[8]Please note that this sequence and allocation of issues is not universally valid.

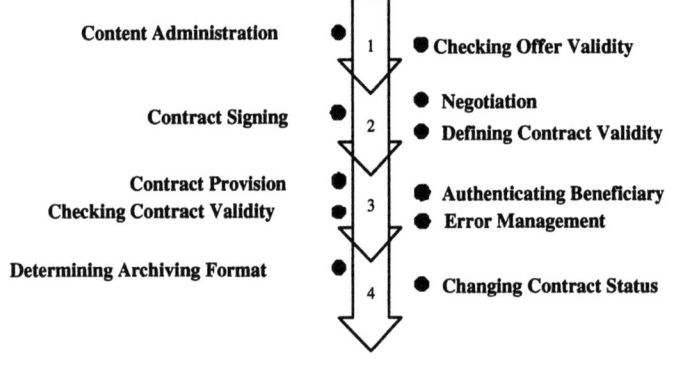

Figure 4.16: Sample operations when managing electronic contracts

Content Administration

Creating an offer implies that the sellers have access to a list of their content. This list should be sorted by sensible criteria, such as format, content type, offer–already–created, etc. For resources on this list, sellers are allowed to formulate offers. Offers should also be visible to the seller after they have been formulated. To enable these features, metadata that describes the content has to comprise the content criteria, and sellers have to be related to *their* content and to the offers *they* have created.

Checking Offer Validity

Checking whether an offer is valid can comprise several activities and varies depending on the concrete application. By all means the digital signature of the offer has to be checked.

- Digital signatures provide several services, e.g. unique identification of the offer signer. Generally, digital signatures are a technical means to provide authentication, verification, non-repudiation and integrity [Sch79]. The offer is only valid if the signature on the offer is verifiable and unambiguously identifies the signer. Along with the signature, sufficient information about the signature such as hash algorithm or key length algorithm needs to be provided to technically facilitate the signature verification.

- Besides verifying the signature it should also be checked whether the contract content is sensible, e.g. if the offered goods are in stock, if the terms and conditions are reasonable, etc.

- After the offer has been signed, the signature unambiguously identifies the seller. Here it is an important management issue to check if the signer is a person that has the authorisation to create and sign an offer. For the offer to be valid, the signer needs to be a person who owns property rights of the offered resource. Often the content creator is such an authorised person, but sometimes retailers or second-hand vendors have such rights as well.

Managing Contract Conclusion

In the contract conclusion phase of the contract life cycle the three issues *contract negotiation, contract signing* and *defining contract validity* play an important role.

Negotiation

On the way to reaching an agreement between buyer and seller concerning terms and conditions of the contract, often a negotiation between contract parties takes place. For example, contract party A offers certain conditions to party B. B has the choice between agreeing and rejecting these conditions. If the conditions are rejected, party B may state new conditions and present them to party A. This process is called *negotiation* (depicted in Figure 4.2) and needs to be technically facilitated in the conclusion phase of electronic contracts.

Contract Signing

After the contract parties have agreed on the contract content, both parties have to sign the offer. At this stage, one party has already signed the electronic offer to make it legally valid. The subsequent signature of the accepting contract party then is called *offer confirmation* and automatically results in the conclusion of the contract. Like the offer signing, the contract signing requires the respective environment. To facilitate digital signatures and their verification in general, a public key infrastructure (PKI) has to be provided. PKI [KL89] offers means to issue key pairs, issue and revoke certificates, create and verify signatures, etc.

Defining Contract Validity

"When is a contract valid?" is a questions of great importance when processing electronic contracts and brings up a lot of crucial issues: Is the contract valid because all contract parties have signed it? How long is a contract valid? Is the contract still valid after it has been processed by an access control service? How can *double spending* of electronic contracts be avoided? Guidelines for these issues should be well–defined in the contract conclusion phase. These guidelines for contract validity have two main dimensions: there are guidelines to make the contract legally binding, and others to make it technically valid.

- *Legal requirements.* If an electronic contract is designed to be legally binding, the signing certificates have to be *qualified*, and the electronic signatures have to meet the nationally required security level [Eur99]. Legally binding electronic signatures have to

    - be exclusively assigned to the signer,
    - unambiguously identify the signer,
    - be created with means that can be kept under the signer's control,
    - ensure integrity of the signed data, and
    - be based on a qualified certificate that has been created by technical components and practices that meet the security requirements of the national law and its regulations.

    A qualified certificate can only be issued by a qualified certification authority. Qualified certification authorities are nationally registered authorities that have undergone a quality and security check by the regulatory authority for electronic signatures.

- *Technical requirements.* First of all, some technical requirements are predetermined by the legal requirements. The system that processes electronic contracts has to provide technology that meets the legal requirements. Besides the implementation of legal requirements a DRM system usually has additional specifications for valid contracts, such as

    - the electronic contracts have to be written in a specific rights expression language (see Chapter 3),

- the contracts need to comprise certain information for specific usage scenarios (see Section 4.5),
- the objects (parties, resources, etc.) in electronic contracts have to be uniquely identifiable by the system,
- the resources stated in the contract have to be available in the system,
- the electronic contract must not have been revoked,
- the electronic contract must not be expired,
- if the electronic contract shall be processed by an access control service, it must not have been processed successfully before (see *check contract status* in Section 4.8).

Managing Contract Fulfillment

This section addresses management issues in the contract fulfillment phase of the contract life cycle, such as *Contract Provision, Checking Contract Validity, Authenticating Beneficiary* and *Error Management*, leaving out the technical issues, such as contract parsing, interpretation and processing.

Contract Provision

The contract fulfillment phase includes the execution of rights. First of all, the contract that shall be executed has to be provided to the executing platform. There are two basic possibilities to provide contracts to a platform:

1. *Contract remains with platform.* After the conclusion of a contract the platform stores and administers the contract. Beneficiaries arrive at executing their rights by authenticating themselves to the platform and then receive access to resources and services according to their contracts.

2. *Ticket/Contract remains with the consumer.* The consumers receive the contract after its conclusion and are required to provide the contract at the time they want to execute their rights.

In the second case, the following management issue arises: if several contracts are in the contract repository of the consumer, who chooses the right contract? Does the consumer need to know what the appropriate contract is, like it is the case with x509 certificates? Or does the platform have the intelligence to pick the right and valid contract from the contract repository of the customer? This example names management issues that will arise with the dispersion and frequent use of electronic contracts.

Checking Contract Validity

An electronic contract is valid if it meets the legal and technical requirements defined in the contract conclusion phase. Technical requirements include the implementation of the legal requirements and the definition of system–specific respectively application–specific requirements. For example, checking contract validity may comprise the following steps: checking the contract status, checking the contract signatures, identifying contract objects, and checking resource availability.

- *Checking Contract Status.* As soon as the electronic contracts arrive at the platform with the request to execute certain rights, the contract processing platform has to check the contract status. This activity includes the inspection of specific contract revocation lists in order to identify expired and legally revoked (i.e. invalid) contracts and to prevent duplication, respectively "double spending" (see e.g. [MN93]), of digital contracts and the herein granted permissions. For example, the "double spending" prevention procedure is necessary if a contract defines a maximum number of uses for certain digital goods/services and the contract has already been executed on the same or on another platform.

  To the best of my knowledge, there are no existing revocation lists for electronic contracts yet. Such revocation lists, would have similar functions as certificate revocation lists [KL89], such as identifying and publishing expired, legally invalid, manipulated and forged contracts, as well as publishing contracts that have already been fulfilled. Running contract revocation lists requires the unique identification of each electronic contract in a closed system.

- *Checking Contract Signatures.* Before a contract is ready for processing, this step verifies the digital signatures of the corresponding contract to ensure the integrity of the contract and the authenticity of the contract parties. This activity comprises all legally and technically required tasks as defined in the contract conclusion phase. In this context the sustainability of electronic contracts is an important issue.

As mentioned earlier electronic contracts signed with "qualified" certificates are legally binding. Digital certificates get a time stamp from the qualified certification authority at the time they are issued and have a certain validity period. After the time period has expired the digital certificate is invalid and no longer suitable for the verification of electronic signatures. To ensure the sustainability and verifiability of digital certificates and thus of electronic contracts, additional time stamp and naming services have to be applied. These services extend the basic functions of digital signatures. The time stamp service affirms that a particular contract has been presented to the time stamp issuer at a certain point in time. The naming service (of a certification authority) handles the binding of signature keys to persons and confirms whether a certificate has been valid at a certain point in time [BRB99]. To verify a signature from a certificate that has expired, the naming service has to be questioned about wether the signature on the respective contract was valid at the time of the time stamp. Thus the verifiability of digital signatures can be provided. This issue is also addressed in the work of Anagnostopoulos et al. [AGT01], which presents data structures that can support an infrastructure for persistent authenticated dictionaries. Applications include credential and certificate validation checking in the past (as in digital signatures for electronic contracts), digital receipts, and electronic tickets.

- *Identifying Contract Objects.* To execute a contract as it is intended by the contracting parties, all objects that occur in electronic contracts have to be uniquely identified. Contract objects are for example, resources, contract parties, etc. Here it is important to distinguish between *system–wide* and *world–wide*, i.e. globally unique identifiers. For some applications it might be satisfactory to uniquely identify the contract objects system–wide. For example, it is adequate to identify subscribers of an online newspaper only by their id, which is unique on the online newspaper platform. In cases where contracts state the royalties for authors and the contracts circulate on various different platforms, an id is required that is globally unique. System identifiers are generated and assigned to objects by the respective system, whereas global identifiers are issued and assigned to objects by international initiatives. Examples of globally unique identifiers are the digital object identifier (DOI) [Nat00] for the identification of digital goods respectively resources and the uniform resource name (URN)

[BLFM98, Moa97] that globally and uniquely identifies a resource or unit of information independent on its location. Also an IP address [Pos81a] could serve for the identification of a certain resource, or the X.500 distinguished name [IT93a] for the identification of a certain individual.

- *Checking Resource Availability.* Prior to finally execute the electronic contract, the system has to check on resource availability. Although the contract can be valid otherwise, it can not be executed if the resource, identified in the contract, is not available.

Authenticating Beneficiary

As referred to in the issue *contract provision* an authorisation of the beneficiary is required for usage scenarios, such as access control. If the electronic contract shall be used to grant access to electronic resources, the access control service first and inevitably demands for the authentication of parties. The system initiates the authentication process for the party that has triggered a particular access request. Possible authentication mechanism are e.g. a Kerberos–based service [SNS88], or an authentication infrastructure based on X.509–certificates [IT93b]. Authenticating the beneficiary is relevant for access control services but not for other purposes, such as accounting.

Error Management

From the previous paragraph one can conclude that electronic contracts are sometimes not valid or not qualified to be processed in certain software services. In cases where electronic contracts have been rejected to be processed, in certain software services an error management needs to be available. The error management defines how to proceed with rejected or unemployable electronic contracts. Like in programming environments, the error management should distinguish between *error severity, error types*, and support *error routines*.

- *Error Severity.* The error management should distinguish different grades of severity of errors. Some errors might be severe and cause the rejection of the electronic contract, while others might not result in the contract rejection but only throw a warning during processing. Errors might be assessed as variably severe depending on the software service that processes the contract. For example, the accounting service rejects a contract if the contract does not include a monetary obligation. As the access control service is not dependent on this information, the missing attribute does not even cause a warning.

- *Error Types.* The error management should distinguish various error types, for example *invalid–signature, contract–expired, contract–already–executed, object–identification–failed, contract–format–invalid, resources–not–available,* etc. The identification of error types is important for the further handling of errors. Depending on what error type occurs the respective *error routine* is called.

- *Error Routines.* Error routines define activities that are executed on the occurrence of a certain error type. For example, if the system is checking an access request for a certain resource and the error type *invalid–signature* occurs, the system would reject the access request. The error type *contract–already–executed* would initiate a routine that rejects the access request and additionally adds the electronic contract to a public list of already "spent" contracts.

Managing Contract Archiving

The archiving of electronic contracts brings up the management issues *Changing Contract Status* and *Determining Archiving Format*. These issues deal with the consequences of contract completion and with proper contract storage.

Changing Contract Status

If a contract has been completed and all rights have been executed, the contract should officially get the status *fulfilled*. Ideally the electronic contract is registered on a public list for fulfilled contracts that other DRM systems and platform may access. Such public lists reduce the risk of contract double–spending (see e.g. [MN93]).

Determining Archiving Format

Electronic contracts that have been executed are often archived. However, the national laws on electronic contracts have different statements for the format of electronic contracts and their archiving. In Germany, for example, solely electronic contracts that require to be stated in written form in the non-electronic world have to be archived [Deu01b]. The archiving periods for written and electronic contracts are the same and depend on the contract type, such as purchase, insurance contract, etc. Such electronic contracts have to be readable for all contract parties and also for third parties, such as lawyers. A detailed treatment of legal issues of electronic contracts would go beyond the scope of this thesis. To read more about these issues in Europe, and particularly in Germany, please refer to the following EU-directives: directive of electronic commerce [Eur00], directive on distant contracts [Eur97], and directive on electronic signatures [Eur99], respectively the implementation into national German law Gesetz zum Elektronischen Geschäftsverkehr (EGG) [Deu01a], Fernabsatzgesetz [Deu00a], and Novelle des Signaturgesetzes [Deu00b].

The above-mentioned management issues occur during the life cycle of electronic contracts. I do not claim that the list is complete, but it names important issues that have been discussed during the development of this work. In a concrete application some issues might become obsolete whereas others might have to be added. The management issues heavily depend on the respective usage scenario.

4.9 Related Work

This section discusses other approaches to modelling electronic contracts. Some approaches include an introduction of their underlying data model of electronic contracts [LDF+02, KGV99, GHM00], which most times refer to one special application rather than to a generic model for electronic contracts. In most of the works a (proprietary) XML-based language for the contract serialisation [NCL+03, SDN+00, ZS01] is used or at least proposed [BJPW99].

Service Level Agreements

In the work of Keller et al. [KKL+02] a management architecture for specifying, deploying, monitoring, and enforcing service contracts is proposed to provide a basis for *service level agreements*. A service level agreement is a contract that defines the technical support or business parameters that an application service provider or other IT outsourcing firm offers its clients. The agreement spells out measures for performance (i.e. quality of service

(QoS)) and consequences for failure. Keller at al. define object classes that represent their contract model. Their contract model is tailored to the needs of service level agreements, and thus contains other contract objects than the contract model introduced in this work (see Section 4.6). However, their model does contain the core contract objects that have been defined in Section 4.4.1: provider and customer (i.e. parties), and service (specifying permission and resources), as well as objects that represent the guaranteed service parameters (i.e. constraints). In a later work Keller et al. [LDF+02] motivate the description of service level agreements in a standardised format, namely the web service level agreement (WSLA) language specification. Keller et al. finally demonstrate the application of this language in [KL02] for the dynamic e–Business. In this work and in the language specification Keller et al. explicitly state that the general structure of a service level agreement can be described with three basic object types: parties, service description, and obligations (see Figure 4.17).

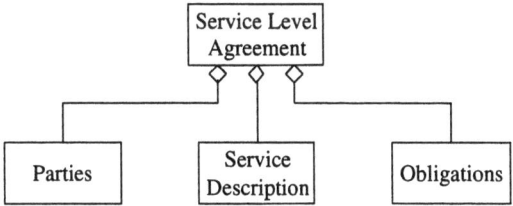

Figure 4.17: General structure of an service level agreement [LDF+02]

Parties are, for example, service provider and service customer. Service descriptions include the guaranteed service (e.g. stock quote service) and the service parameters (e.g. availability and response time). The service parameters can be seen as *rights to a respective service* (i.e. permissions). Obligations define the appropriate actions (duties), e.g. payments to be taken if a violation of a guaranteed service level has been detected. One important difference from the licenses and contracts presented in this thesis is that service level agreements define *not one* but *several different* sets of permissions and duties between the contract parties. All these sets are bound to a condition. When executing the contract one of these conditions is met and the respective set of permissions is granted. For example, the service provider receives a basic fee for a service, e.g. a stock quote service. If the service provider meets all conditions (constraints) that have been defined in the service level agreement, the fee will be regularly paid. If the service provider can not provide the promised conditions (e.g. availability and/or response time) of the service, the payable fee will be reduced (as specified in the agreement). The introduced rights expression languages are not designed to support service level agreements. A subject of future work is an analysis about whether the predominant RELs are or should be able to express service level agreements.

The work of Meredith and Bjorg suggests using electronic contracts to ensure *quality of service* in a technical environment where web services discover and interact with each other [MB03]. The work does not propose a specific rights language but discusses, what such contracts should cover.

Rule XML

The Babel project aims at providing interoperability between inhomogeneous applications [ZS01]. Babel supports the specification of related applications in terms of the functions they deliver and the data they expect as input and produce as output. Furthermore, it enables the specification of business rules for how these functions should be integrated, which can be seen as a kind of agreement. The business rules are formulated in the XML-based language *Rule XML*.

Trading Partner Agreements

Executable Trading Partner Agreements (TPA) [SDN+00] are contracts that trading partners in electronic commerce have agreed on. The agreements are formulated in the XML-based trading partner agreements Markup Language (tpaML) [Sac00]. The language specification has been submitted to OASIS[9]. TPAs additionally contain policy information for different layers in the protocol stack, whereas the contracts addressed in this work only contain information of the application layer. The TPAs contain an agreement on functionality and services that the trading partners offer to each other. Rather than agreeing on usage rights over digital goods the partners agree on predefined and implemented procedures, such as "reserve hotel" that may be called by the remote trading partner.

---

[9]See: Organization for the Advancement of Structured Information Standards, http://www.oasis-open.org/

Contract Aware Components

Beugnard et al. [BJPW99] introduce a general model of software contracts that aims at increasing trust and reliability between software components. To conclude contracts between components, every component publishes a feature set to describe its services in a common language (e.g. CORBA IDL). Contracts are established between a client and server component in a negotiation phase in which the contract parties agree on certain services. The work provides a basic interface description for the negotiation phase. Beugnard et al. suggest an "XML–formatted description of the contracts" that is applied for negotiation (i.e. interoperability) purposes.

Electronic Contracts Used for Workflow Management

Crossflow[10] is a European Community research project into support for cross–organisational workflow management in virtual enterprises. In the project electronic contracts are used to define transactions between automated systems [KGV00]. The automated systems on both sides are Workflow Management Systems (WFMSs), extended with contract handling facility that supports outsourcing interaction. On the one side there is a consumer WFMS that desires to outsource a process, on the other side there is a provider WFMS that is capable of executing this process for the consumer.

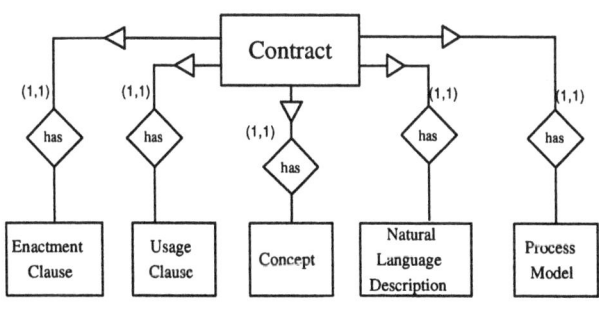

Figure 4.18: Simplified model of contracts applied in a WFWM [KGV99]

The contracts that are used in this project are designed to be deployed in a workflow management application. Therefore, the five basic contract

---
[10]See: http://www.crossflow.org/

elements that are identified describe all aspects of automatic workflow (see Figure 4.18): the *concept model* defines and assigns values to all objects that can be parameterised in the contract; the *process model* describes the internal structure of the workflow process implementing the service; the *enactment model*, describing details of the enactment; the *usage model*, defining manners in which the contract can be used; the *natural language description*, being a piece of text that is not meant for electronic interpretation but describes the service in an easily understandable way and refers to the legal context of the transaction. These basic contract element differ greatly from the general core contract objects identified in Section 4.4.1, because they serve one specific application. Such specialised contracts are hardly applicable in other usage scenarios (see Section 4.4.2). Within the scope of the same project the states of such workflow contracts have been identified [KGV99]. The identified states are: contract template, contract advertising template, contract search template, contract instance, and partially filled contract. These states occur in the first two phases of the contract life cycle, defined in Section 4.1; contract execution and contract archiving are not addressed in detail.

Legal Aspects of Electronic Contracts

The work of Gisler et al. [GSSG00] considers the business phases and the legal phases of digital contracts. Considering the precedent actions between the later contract parties and not addressing the archiving of electronic contracts, their model results in four phases that differ from the approach presented in this thesis: *Information Phase* (Contract Conception), *Intention Phase* (Contract Preparation), *Agreement Phase* (Contract Negotiation), and *Settlement Phase* (Contract Fulfillment). The first two phases of their model do not represent contract states that are legally relevant, and the contract archiving is not considered. Therefore, I disagree with contract life cycle approach.

Business Contracts

In a very early work Milosevic and Bond make an initial attempt to address electronic contract issues [MB95]. They introduce a generic business contract framework comprising the issues *Contract Domain, Contract Template, Contract Negotiation, Contract Validity, Contract Monitoring*, and *Contract Enforcement*, and apply these measures to contracts on the Internet. As basic elements in contract templates they identify the *roles* of the parties, the *period* of the contract, the nature of consideration (*resource*), the *obligations* associated with each role, and the *domain* of the contract. The generic business contract framework of Milosevic and Bond comprises legal and technical issues which they do not clearly distinguish. The basic contract elements are not described in detail and differ from the approach shown in this thesis (see Section 4.4.1). Roles are considered as *one* means for party identification, and the period is *one* of the several relevant contract attributes. In contrast to this approach I regard the contract domain as something that influences the contract content but is not part of the contract.

In a later work Milosevic et al. address *Business Contracts for B2B* [GHM00]. Here, they redefine the basic elements of business contracts, name typical contract phases in the B2B business, and present some implementation experiences. In this work a contract modelling approach is introduced that allows for the contract components: a preamble (involved parties), an approval section (enumerating those who have approved the contract), digital signatures, contract clauses, and policy specifications. This approach differs from my approach by formulating the entire basic contract information in the policy specification. The policy specification comprises the contract party information, contract permissions, duties, and constraints, and relates this information to programming logic (instead of a markup language). The formulation of the policy specifications has been influenced by the Event–Condition–Action paradigm from active databases and the ODP enterprise language. The latest work of Milosevic et al. introduces the requirements for a *Business Contract Language* [NCL$^+$03] and addresses discretionary enforcement of electronic contracts [MJP02].

Chapter 5

Design of a Rights Expression Exchange Framework

This thesis has the objective to develop methods and tools that support the exchange and processing of rights expressions (see Section1.3). Rights expressions are valid instances of a rights expression language (see Section 3.1). Offers, contracts, and licenses formulated in a REL are specific rights expressions, i.e. instances with certain semantics in the contract life cycle (see Section 4.1). Whereas Chapter 4 has covered the contract (respectively rights expression) content, and its tailored composition that support later *processing*, this chapter focuses on the *exchange practices* of rights expressions. The first section introduces a general communication model and deploys its principles to the exchange of rights expressions (Section 5.1) which results in the *rights expression communication model*. A technical design for the implementation of such a rights expression communication model – the *rights expression exchange framework* – is presented in Section 5.2. Here, the framework's functional and technical perspectives (see Section 2.2.2) are discussed in detail.

5.1 Exchanging Rights Expressions

In the following sections an outline is given of the exchange of rights expressions between sender and receiver. Therefore, Section 5.1.1 introduces the general communication model, which is later adopted to the communication via rights expressions (see Section 5.1.2). Section 5.2.1 describes the component-based technical design of a rights expression exchange framework. Finally, in Section 5.2.2, a check list of technical requirements for such a rights expression exchange framework is provided.

5.1.1 The Communication Model

Basic components of communication can be found in the communication model [Sch71] of Schramm illustrated in Figure 5.1. According to Schramm, *communication* is the the sharing of an orientation toward a set of informational signs ... where both source and destination must share a field of experience. Constituents of Schramm's communication model are the source, the encoder, the signal, the decoder, and the destination.

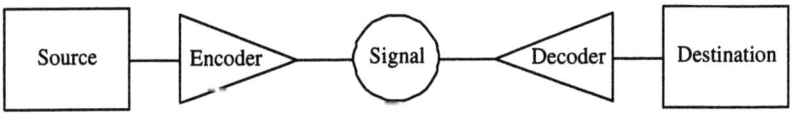

Figure 5.1: The communication model [Sch71]

- *Source.* According to Schramm's model, a message begins at a source. The source can be a person or computer system that aims at sending a message to a destination, where a person or a computer receives this message.

- *Encoder.* The communication encoder is responsible for taking the ideas of the source and putting them in code, expressing the source's information in the form of a message. Therefore, the encoder has to choose a coding language that is understandable for the destination.

- *Signal.* The signal is the encoded message, an instance of a language that source and destination are familiar with, which is sent through a communication channel towards the destination.

- *Decoder.* Like a source needs an encoder to translate the original message into a transportable message, the destination needs a decoder to retranslate. To decode the message the decoder needs to understand the language in which the signal has been coded. After decoding, the message is identical with the earlier-sent original.

- *Destination.* For a successful communication, there must be somebody at the other end of the channel. This person or computer system can be called the destination. After decoding, the destination is able to understand and process the message, and possibly send a reply on the message.

Schramm's communication model describes communication in general. In the next section, Schramm's model is deployed to the exchange of rights expressions between two computer systems.

5.1.2 The Rights Expression Communication Model

When exchanging rights expressions the communication between two cooperating computer systems corresponds to the basic communication model. In this section the communication model is adopted to the communication via rights expressions between two DRM system components, resulting in the *rights expression communication model*. In the following, the required steps are identified which facilitate the exchange of rights expressions between two computer systems. It is assumed that after exchanging the rights expressions, each system is capable of processing them internally, for example, in an access control service context, in which permissions are assigned to parties or in an accounting service context, relating parties to monetary duties.

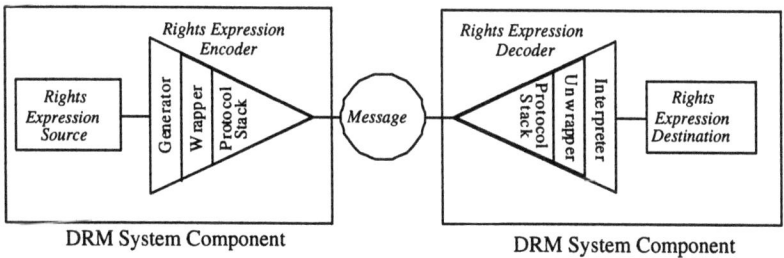

Figure 5.2: The rights expression communication model

The rights expression communication model has five stages: the rights expression source, the rights expression encoder, the message in form of a rights expression, the rights expression decoder, and the rights expression destination (see Figure 5.2). The communication takes place between two

DRM systems or DRM system components (see Section 2.3). A DRM component in this case is a DRM subsystem, such as the DRM platform or the DRM client (e.g. secure viewer).

- *The Rights Expression Source.* In the rights expression communication model a DRM system or a person is the origin of the the rights information to be delivered, i.e. the rights expression source.

- *The Rights Expression Encoder.* This stage provides three tasks: the message generation, the message wrapping, and the application of the respective protocol stack:

    - *The Generator.* The generator supports the coding of the rights expression source into a rights expression. The rights expression can either be generated by the DRM system or be provided by a human actor (e.g. the content provider). The rights expression is formulated in a rights expression language (see Chapter 3) that has to be well–known by the receiving DRM component in order to be understood

    - *The Rights Expression Wrapper.* As rights expressions comprise sensitive information, they are usually not transmitted in plain text and/or without protection. The rights expression wrapper "wraps" the rights expression and thus provides confidentiality and integrity of the rights expression, as well as authentication and verification of the rights expression issuer or signer. Here, the term "wrapper" is not to be understood as the wrapper or object adapter, described in [GHVJ94].

    - *The Protocol Stack Application.* Referring to the TCP/IP protocol stack, the generator and wrapper are encodings on the application layer. To be sent over a network (e.g. the Internet), the protocols of lower layers, i.e. the transport layer (e.g. TPC [Pos81b] or UDP [Pos80]) and the internet layer (e.g. IP [Pos81a]) have to be applied to the rights expression. The protocol stack to be applied depends on the actual network and on the communication partner. Applying the protocol stack produces the rights expression *message*.

- *The Rights Expression Message.* The rights expression message comprises a wrapped REL instance that is sent to the receiving DRM component via a communication channel, i.e. the rights expression massage is a transport format for the encoded rights expression source.

The message can be transported via any kind of network, e.g. the Internet, a virtual private network (VPN), but also a smart card or floppy disc.

- *The Rights Expression Decoder.* This stage provides three tasks: the reverse application of the protocol stack, the message unwrapping, and the message interpretation:

    - *The Reverse Application of Protocol Stack.* This encoding task makes the wrapped rights expression available for the application layer, i.e. the unwrapper and interpreter.
    - *The Rights Expression Unwrapper.* The rights expression unwrapper is the complementary component to the rights expression wrapper. It unwraps the rights expression. This component also provides for the extrinsic checking of the digital contract, such as checking the contract integrity and the contract signature.
    - *The Interpreter.* The interpreter transforms the encoded rights expression, formulated in the rights expression language, back to the original rights information. This rights information can then be processed in the receiving DRM system component in various ways. The interpreter that has been implemented within the scope of this thesis stores the original information in a generic contract schema that is a means for flexible and open processing of rights information. To read more about the generic contract schema, please refer to Section 4.6.

- *The Rights Expression Destination.* The receiving DRM component is the destination of the rights expression, where it can be processed in various software services (see Section 4.4.2).

5.2 The Rights Expression Exchange Framework

This section provides a general technical design for a rights expression exchange framework and a check list for its implementation. The rights expression exchange framework has been derived from the rights expression communication model defined in Section 5.1.2. Such a framework facilitates the exchange of rights information between two dislocated computer systems. The framework description is independent from a concrete technical implementation.

5.2.1 Technical Design

The rights expression exchange framework presented in this section is a framework that facilitates the exchange of rights information between two or more components, and is based on the rights expression communication model (see Section 5.1.2). The framework design deploys a component-based approach, i.e. a set of cooperating components that offer the necessary functions for rights expression exchange. Each of the components is autonomous and can be reused in other frameworks; new components can be easily added. A rights expression exchange framework is intended as additional module in a web server or application server and can be easily integrated into an existing software environment. In general, we propose a dynamically extensible component framework, as discussed in [GNZ00].

According to the rights expression communication model, the following components that are required for a rights expression exchange framework can be derived: the rights expression generator, the rights expression wrapper, the rights expression unwrapper, the rights expression interpreter, and the mediator. The generator and the wrapper are designed to encode rights expressions; the unwrapper and interpreter are designed to decode rights expressions (see Figure 5.3). Considering the four–layer protocol TCP/IP, the four components are resident on the application layer. It is assumed that the protocol stack below the application layer, such as TCP/IP [Pos81b, Pos81a] is available to the framework. The framework interacts with other software services, such as web servers or application servers, and reuses a data base service for the (temporary) storage of rights expressions.

- *Rights Expression Generator.* The product of the rights expression generator is a schema and specification–conforming instance of a rights expression language. The rights expression generator has to know the syntax and the semantics of each REL it is supposed to create instances of. Coding is necessary for each new REL the generator adopts. Rights expressions can be created by machines or by human actors. Often rights expression, such as offers, licenses, contracts are created by human actors. Therefore, the rights expression generator component can be implemented as graphical user interface that guides the user through the generation process. If rights expressions and in particular electronic contracts undergo a tailored composition (see

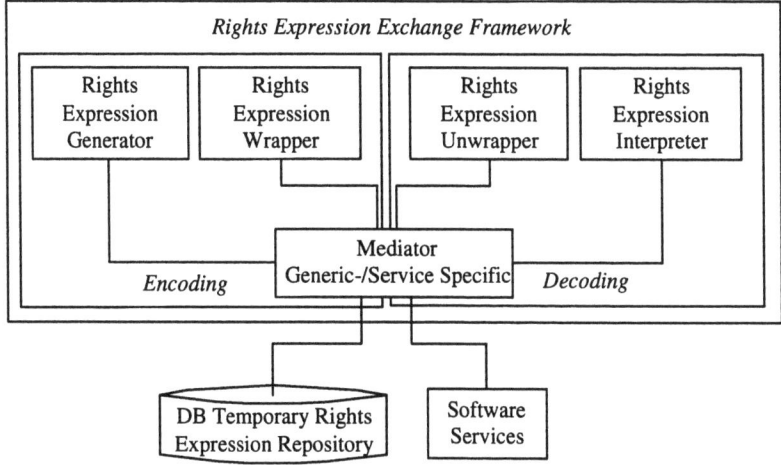

Figure 5.3: Components of a rights expression exchange framework

Section 4.5), the number and type of application–specific objects and attributes are identified. A rights expression generator has to support the creation of such tailored contract templates. Completed rights expressions and rights expression templates can be (temporarily) stored in a database, the *rights expression repository*. The implementation and usage of a rights expression generator is described in Section 6.2 respectively in Section 7.2.2.

- *Rights Expression Wrapper.* The rights expression wrapper component is in charge of a further encoding of the rights expression that has been created by the rights expression generator. The wrapper applies technical means to the rights expression to provides various security services (such as listed in Section 5.1.2). The required security means to be applied (encrypting the rights expression or digitally signing the rights expression) depend on the software service that later processes the rights expression. Thus, the rights expression wrapper has to be customised and possibly extended for its actual application. The wrapper respectively all its methods that perform security applications can be controlled via its application programming interface. The wrapper receives and returns the rights expression from/to the mediator component (see below). The implementation and the usage of a rights expression wrapper is described in Section 6.4 respectively in Section 7.2.3.

- *Rights Expression Unwrapper.* The rights expression unwrapper component is in charge of decoding all applied security means, and thus is the reverse functionality of the wrapper component. Here, the rights expression is decrypted and digital signatures are verified. Again, the security checks that have to be applied depend on the software service that processes the rights expression. Thus, also the rights expression unwrapper has to be customised and possibly extended for its actual application. The methods that the unwrapper provides can be called via its application programming interface. The usage of the unwrapper API is performed by the mediator. The implementation and the usage of a rights expression unwrapper is described in Section 6.4 respectively in Section 7.2.3.

- *Rights Expression Interpreter.* This component is in charge of decoding the rights expressions to the original rights information. The rights expression interpreter has to know the syntax and the semantics of each REL it is supposed to interpret instances of. Coding is necessary for each new REL the interpreter adopts. After the interpretation the rights information should be ready for processing in software services. Therefore, the interpreter maps the rights expression to an application–specific contract schema (see Section 4.6.1). The application–specific CoSa covers all attributes that are required by later software services. The application–specific CoSa is queried by the mediator via the generic CoSa API (see Section 4.6.3).

- *Mediator.* The mediator plays a central role in the rights expression exchange framework: it coordinates or *glues* the functions of the remaining four components in the framework. The mediator interacts with the framework components and other scenario–specific software services, such as access control mechanism or web server. With the knowledge of the APIs of the remaining four framework components, the mediator is able to coordinate them. By coordinating various components via their API, the mediator controls the workflow between the framework and the software services. The mediator code can be implemented as separate software class/program, or integrated in existing software classes/programs. The mediator code is adapted to the needs of a specific application. Very little of the mediator code can be reused in other applications, therefore in most cases each new application that uses the framework components requires a new mediator implementation.

5.2.2 Implementation Check List

The following technical requirements are necessary for the implementation of a rights expression exchange framework:

1. *Rights Expression Language.* For the exchange of rights information a respective language is required. There are a number of rights expression languages, which are usually freely available (see Section 3.4). The cooperating DRM systems should agree on a rights expression languages(s) and on REL application policies.

2. *Implementation of a Generator.* Each DRM component that aims at encoding rights information requires a rights expression generator. If DRM components run on different operating systems or platforms, a generator has to be developed for every operating system, respectively platform. The generator has to support a suitable interface. Offers or contracts that are created by content providers usually require an interface for the manual input of rights, license or contract information. However, there might be applications that require an API for the formulation of rights expressions. Furthermore, a generator might require a database connection for the temporary storage of licenses, because sometimes offers need to be restored for modification or further processing by a different component, e.g. the mediator.

3. *Implementation of a Parser.* A parser is required for all DRM components that receive rights expression messages. A parser usually provides validating the rights expressions in terms of syntactic correctness. A parser has to be available for the different operating systems respectively platforms of the DRM components. For rights expression languages that are XML-based, a large number of XML parser implementations are freely available on the Internet.

4. *Implementation of an Interpreter.* After a rights expression has been parsed, an interpreter maps the rights expression to semantics. The semantics are derived from REL in which the rights expression is coded. Additionally, the interpreter provides the rights information in a processable format. Again, an interpreter has to be available for the different operating systems respectively platforms of the DRM components. In this thesis a rights expression language interpreter for

ODRL (see Section 3.4.1) has been developed that is freely available in the spirit of open source.

The development of rights expression language interpreters is at the very beginning. Apart from the work at hand, no design or comprehensive implementation of a rights expression language interpreter is available.

5. *Secure Transportation Channel.* The exchange of a rights expression requires a secure transportation channel between two appointed DRM components. The channel has to assure that the rights expression is not deleted or modified during transportation. With the adequate security provided, the Internet or private networks are potential transportation channels.

6. *Concept for Runtime Presentation.* After the message has been interpreted the rights information is processed in software services. A general runtime representation of the rights information has the advantage that it is suitable to serve various software services, such as access control, accounting, and CRM services. The information in such a general representation is usually accessed via a predefined application programming interface. A generic schema for the representation of rights information as objects has been introduced in Section 4.6.

7. *Temporary Storage.* Rights expressions, in particular licenses and contracts, undergo several phases in their life cycle (see Section 4.1). Very often, such right expressions need to be temporarily stored. Depending on the current phase, different types of storage have to be provided. For example, prior to contract conclusion it is reasonable to store a contract in software objects, an XML database, or a relational database rather than as an XML document in the respective rights expression language, because it might be easier to restore the contract information for modifications in the negotiation phase.

8. *Framework Integration.* Most systems have not been designed from the beginning to handle rights expressions. A rights expression exchange framework often has to be integrated belatedly into application servers. To enable an easy integration, the framework respectively all its components should be coded in a suitable programming language and provide a well–defined interface.

9. *Mediator Implementation.* The mediator does not provide a general functionality, like the contract wrapper, unwrapper or the interpreter. It combines given components for a specific usage scenario. Therefore, a new mediator has to be implemented for each usage scenario. There are mediator tasks that need to be performed in every usage scenario, such as contract unwrapping and interpretation which follow a fixed pattern. In contrast to these, scenario–specific mediator tasks are differing in each application. A scenario–specific mediator task, for example, implements all rights expressions from a given contract into an access control mechanism. Another example of a scenario–specific mediator task is gathering the purchased products from all concluded contracts for the purpose of customer relationship management.

10. *Security Means.* To establish a reliable rights expression exchange framework, a large number of security means have to be applied. Attacks on rights expressions and their reliable processing can have various facets. Attacks that are concentrated on the rights expression message should be fended by the contract wrapper, respectively unwrapper. However, this provides neither yet tamper resistance of the framework components nor prevention from internal attacks (e.g., personnel). Certainly, building real–world tamper resistant systems is a complex task especially in open, distributed environments (see also [LTM$^+$00, AK96])

Chapter 6

Implementing the Rights Expression Exchange Framework

The design steps of a rights expression exchange framework has been introduced in Chapter 5. Originally derived from the basic communication model (see Section 5.1.1), and adapted to the exchange of rights expressions, the technical design of a rights expression exchange framework comprises five components: the rights expression generator, wrapper, unwrapper,interpreter, and mediator. A technical implementation of these components will be described in this section. Section 6.1 presents the underlying software architecture of the framework. The subsequent chapters describe the implementation details of the five components (generator see Section 6.2, interpreter see Section 6.3, wrapper and unwrapper see Section 6.4, and mediator see Section 6.5). Each of the components is implemented in order to be used autonomously. The section concludes with related work in the field of rights expression exchange, i.e. existing implementations of REL interpreters or frameworks (see Section 6.7) .

6.1 Software Architecture

Figure 6.1 shows the software architecture of the rights expression exchange framework. All components of the rights expression exchange framework (i.e. the generator, the wrapper, the unwrapper, the interpreter, and the mediator) are coded in XOTcl (see Section 6.1.1). XOTcl is a suitable language for the implementation due to its strength as *glue code* between components. Additionally XOTcl facilitates the reuse of tools, such as OpenSSL, tDOM, and MySQL.

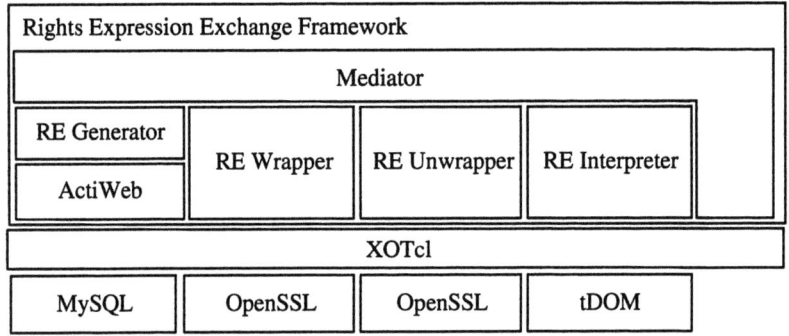

Figure 6.1: Technology used in the rights expression exchange framework

OpenSSL (see Section 6.1.5) as cryptographic library is used in the wrapper– respectively unwrapper component, whereas the generator component is supported by MySQL (see Section 6.1.4) and also uses the ActiWeb package – a package implemented in XOTcl providing web technologies. The rights expression interpreter uses tDOM, an open source XML processor, which implements the Document Object Model (see Section 6.1.3) and allows the handling of XML documents. The mediator is a pure XOTcl implementation and is not reusing other software packages. The mediator has a central function in the framework: it coordinates the remaining four components. The subsequent sections describe the reused tools in more detail.

6.1.1 The XOTcl Language

XOTcl (eXtended Object Tcl) [NZ00b] is an object–oriented scripting language developed by Gustaf Neumann and Uwe Zdun, and which is based on Object Tcl (OTcl) [WL95]. OTcl itself extends Tcl (Tool command language) [Ous94] and provides for the following features: encapsulation, (multiple) inheritance, method chaining, meta–classes, read/write introspection, and dynamic extensibility. XOTcl adds further functionality that helps building and managing complex systems, namely dynamic aggregation, nested classes, assertions, meta–data, per–object mixins, per–class mixins, filters, and dynamic component loading[1] (see Figure 6.2).

XOTcl can be loaded into every Tcl–compatible environment, such as *tclsh* or *wish*, and can be embedded in C programs. XOTcl optionally provides its own shell, called *xotclsh* which is a *tclsh* with XOTcl preloaded. This shell takes the commands (similar to *tclsh*) one by one from a file or from the console. For that reason all Tcl commands remain available and are also applicable on the extension constructs. As Tcl is equipped with appropriate functions for the easy gluing of components, it is a well suited language for the implementation of the components, in particular the mediator component, of a rights expression exchange framework.

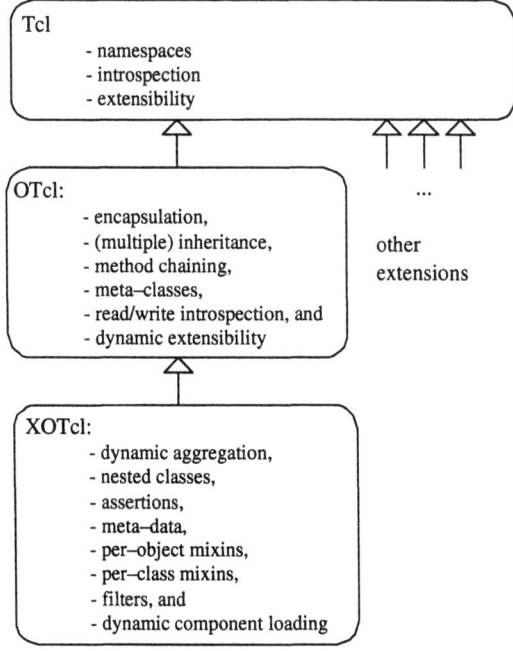

Figure 6.2: Features of XOTcl, OTcl, and Tcl

---

[1]See also: http://www.xotcl.org/

6.1.2 ActiWeb

ActiWeb is a open source library of components for Internet applications written in XOTcl (see Section 6.1.1), including implementations of HTTP servers, HTTP clients, and mobile code [NZ00a], and the support to process common internet formats (e.g. HTML or XML). The main components of ActiWeb are (see Figure 6.3) :

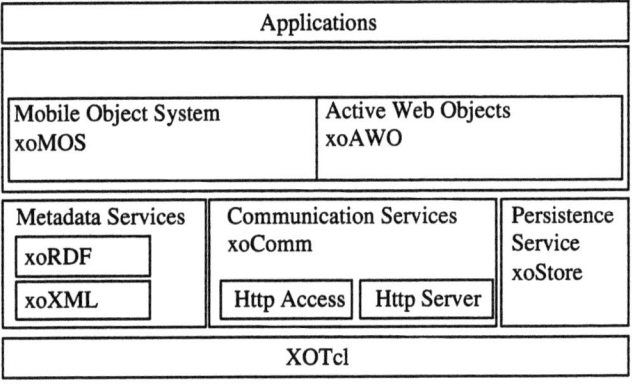

Figure 6.3: Basic architecture of ActiWeb [NZ00a]

- *xoComm*. This component is an object–oriented, highly flexible and configurable HTTP/1.0 [BLFF96] and HTTP/1.1 [FGM+99] server and client access implementation. It supports client access to several other common web protocols. Due to the component–based implementation approach of ActiWeb, compatible components can be substituted, e.g. xoComm can be exchanged for another HTTP implementation providing the same interface.

- *xoXML*. This object–oriented implementation in XOTcl is part of the metadata services supported by ActiWeb. It provides an environment for parsing XML–files into an abstract syntax tree and vice versa.

- *xoRDF*. This implementation is the second of currently two metadata services supported by ActiWeb. xoRDF is an XML/RDF[2] parser/ interpreter environment for XOTcl.

---
[2]See: http://www.w3.org/RDF/

- *xoStore.* This implementation provides a general, persistent store for XOTcl objects which makes objects and their data transparently persistent. xoStore is a wrapper for several database implementations. The version in the current XOTcl distribution is based on several free storage systems, such as GDBM[3].

- *xoMOS.* This component implements a mobile object system based on HTTP and RDF.

- *xoAWO.* The XOTcl implementation of an active web object system allows to easily build web applications with active web documents and web facades to agents. For example, xoAWO enables an agent to use a web representation, such as HTML.

For the implementation of the rights expression exchange generator the xoComm component of XOTcl is used, because it is freely available and provides all web server functionality that is needed to implement a web–based user interface for the generation of rights expressions. A detailed description of the integration of xoComm is given in Section 6.2.

6.1.3 Document Object Model (DOM) Implementation

The World Wide Web Consortium[4] states that the

"... Document Object Model (DOM) is a standard Application Programming Interface (API) to the structure of documents; it aims at making it easy for programmers to access components and to delete, add, or edit their content, attributes and style ... The Document Object Model is a platform- and language-neutral interface providing a standard set of objects for representing HTML and XML documents, a standard model of how these objects can be combined, and a standard interface for accessing and manipulating them [Wor00]."

The document object model (DOM) can be used to process XML documents. It presents all elements of static XML (or HTML) documents as a hierarchy of *Node* objects, such as *Document, Element, Text, Comment* or *Entity*. Some types of nodes may have child nodes of various types (e.g. Document, Entity, and Element), whereas others are leaf nodes that cannot have anything below them in the document structure (e.g. Text and

---
[3]See: http://www.gnu.org/software/gdbm/
[4]See: http://www.w3.org/

Comment). The representation of document objects results in an object tree, also called DOM–tree that starts with a Document (root) element(see Figure 6.4).

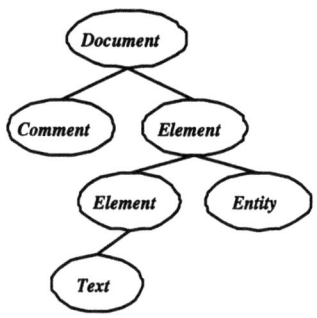

Figure 6.4: A general DOM–tree

To process an XML–based document, for example an electronic contract written in ODRL, the DOM-tree of such a contract contains all elements that are written in the contract instance represented by individual objects. The DOM-tree can than be easily queried or modified via the DOM API or via XML–specific query language such as XPath [CD99] for example. The DOM API contains methods such as hasAttribute(..), getElementsByTagName(..), getTagName(..), etc., with which elements (or tags) of an XML document can be located and analysed.

DOM is designed at several levels:

- Level 1: This level concentrates on the actual core, HTML, and XML document models. It contains functionality for document navigation and manipulation.

- Level 2: This level includes a style sheet object model, and defines functionality for manipulating the style information attached to a document. It also enables traversals on the document, defines an event model, and provides support for XML namespaces.

- Level 3: This level will address document loading and saving, as well as content models (such as DTDs and XML schemata) with document validation support. In addition, it will also address document views and formatting, key events, and event groups. Public working drafts, candidate recommendations, and one recommendation (validation) are available at the W3C[5].

tDOM[6] is an open source implementation in C of the document object model, providing a DOM binding to the Tcl language and thus also to XOTcl (see Section 6.1.1). The most recent tDOM release supports DOM Level 2 and comprises the latest version of *Expat*, the XML parser of James Clark[7], including namespace and DTD support. In this thesis tDOM is used by the rights expression interpreter implementation to parse and access electronic contracts respectively rights expressions written in an XML–based rights expression language. tDOM was chosen because it is freely available and provides a DOM Level 2 implementation which comprises all required functions for processing XML-based documents.

6.1.4 MySQL

The MySQL database server is a fairly popular open–source relational database management system (RDBMS) that uses Structured Query Language (SQL). SQL is a popular language for adding, accessing, and processing data in a database. It is fully multi–threaded using kernel threads, provides API for C, C++, Eiffel, Java, Perl, PHP, Python, and Tcl, allows for many column types, and offers full operator and function support in the SELECT and WHERE parts of queries. MySQL currently runs on the Linux, Unix, and Windows platforms.

In this thesis the MySQL database server distribution *mysqltcl* version 2.40 has been used[8]. It has been applied in the implementation of the rights expression generator (see Section 6.2) for the (temporary) storage of generated rights expressions respectively electronic contracts. MySQL was chosen because of its free availability and flexible nature; furthermore it provides fast reliable database services.

6.1.5 OpenSSL

The Secure Sockets Layer (SSL) is a commonly–used protocol for managing the security of a message transmission on the Internet. In 1999, SSL

---
[5]See: http://www.w3.org/DOM/
[6]See: http://www.tdom.org/
[7]See: http://www.jclark.com/xml/
[8]See: http://www.xdobry.de/mysqltcl/

was succeeded by the open specification of Transport Layer Security (TLS) protocol [DA99], which is based on SSL. SSL/TLS provides for the security aspects *integrity, confidentiality,* and *authenticity* when transmitting messages over the Internet.

SSL/TLS uses a program layer between the application layer and the the transport layer, i.e. SSL/TLS is located on top of the Transmission Control Protocol (TCP) and therefore provides its services for application protocols, such as HTTP, FTP, Telnet, etc. SSL was the de facto standard until evolving into Transport Layer Security. SSL/TLS uses symmetric and asymmetric cryptography and is an integral part of most web browsers (clients) and web servers. If both web client and server support SSL/TLS, they can perform the "SSL-Handshake" to establish a "secure channel". The secure data transmission is then handled autonomously between client and server.

The high availability of SSL/TLS in common browsers has been achieved by the open specification of SSL/TLS and a large number of open–source implementations. The OpenSSL Project[9] is a collaborative effort to develop a robust, commercial–grade, full–featured, and open–source toolkit implementing the SSL/TLS protocols as well as a general purpose cryptography library. The project is managed by a worldwide community of volunteers that are developing the OpenSSL toolkit and its documentation. OpenSSL is used in this project for the implementation of the rights expression wrapper, respectively unwrapper, e.g. the wrapper uses a hash function and asymmetric encryption to create digital signatures of rights expressions. The unwrapper uses the same functions to verify digital signatures. OpenSSL has been chosen because it is freely available and widely-used, and because ActiWeb uses OpenSSL to establish secure HTTP connections.

6.2 The Rights Expression Generator

The generator is a part of the rights expression encoder (see rights expression communication model in Section 5.1.2). It transforms the original rights information that has been provided by DRM system component or a human actor (e.g. a contract party) into a rights expression (see Section 5.1.2). This rights expression is formulated in a rights expression language (see Chapter 3). Therefore, the rights expression generator has to adopt the syntax and semantics of a REL. The generator developed within this thesis,

---

[9]See: http://www.openssl.org/

with the kind assistance of Margit De Toma, is capable of transforming rights information into instances of the open digital rights language (ODRL) version 1.1 (see Section 3.4).

6.2.1 Functional Description

The rights expression generator is a web–based tool that provides among other things a graphical user interface to formulate rights expressions (e.g. offers, licenses, or contracts). The beneficiaries of this tool are e.g. contracting parties that desire to state terms and conditions in the form of an electronic contract. The web–based rights expression generator guides the contracting parties through the process of contract creation. The process starts with supporting rights holders to select a resource that is owned and/ or controlled by them. For this respective resource an offer can be created. This offer usually comprises usage or access permissions and their terms and conditions for the specific resource. Figure 6.5 shows the generator GUI that offers the user the set of ODRL tags (e.g. offer, agreement, asset, permission, etc.) that may be attached to the ODRL root element `rights` (displayed with the term *Rights expression* in Figure 6.5 on the left). Each attached ODRL element may have further sub elements. Each ODRL element selected by the user is then added to an XML document. The document with its tree structure is presented on the left hand side of the tool. Depending on the element type that has been chosen in the tree, the top menu changes and offers all sub elements that may be attached to that particular element (according to ODRL version 1.1). Some elements, e.g. *name, id,* etc. of the context element, require extra input of actual values, such as strings or integers.

The data that is provided by the user of the generator tool is build into a rights expression which is compliant to ODRL version 1.1. When the option *Show xmlfile* is clicked the ODRL instance is displayed in a separate window (see Figure 6.6). The storage location of the XML contracts can be customised according to requirements of the actual application. Clicking on *Store Rights Expression in the database* activates the storage of the rights expression in the underlying MySQL data base in a format that resembles the ODRL document structure. The storage in the data base facilitates the reconstruction of the rights expression at any time. A reconstruction is required if, e.g. if the generator tool shall support contract negotiation (see Section 4.2) and the contracting parties desire to modify an offer until an agreement has been found. Each user is able reaccess and delete his/her

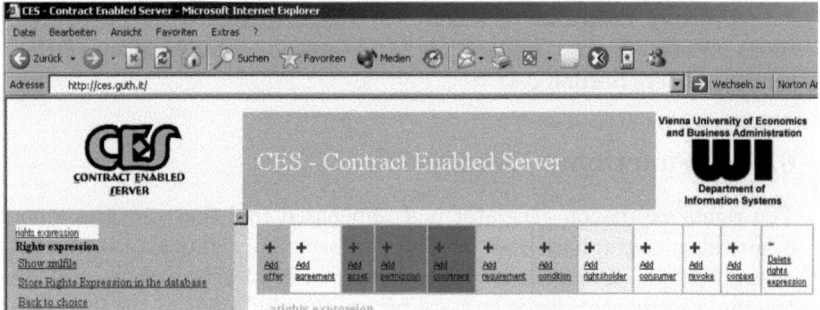

Figure 6.5: Choice of ODRL tags

created rights expressions, i.e. if a content provider has created offers to his/her resources, s/he is able to modify or delete those offers.

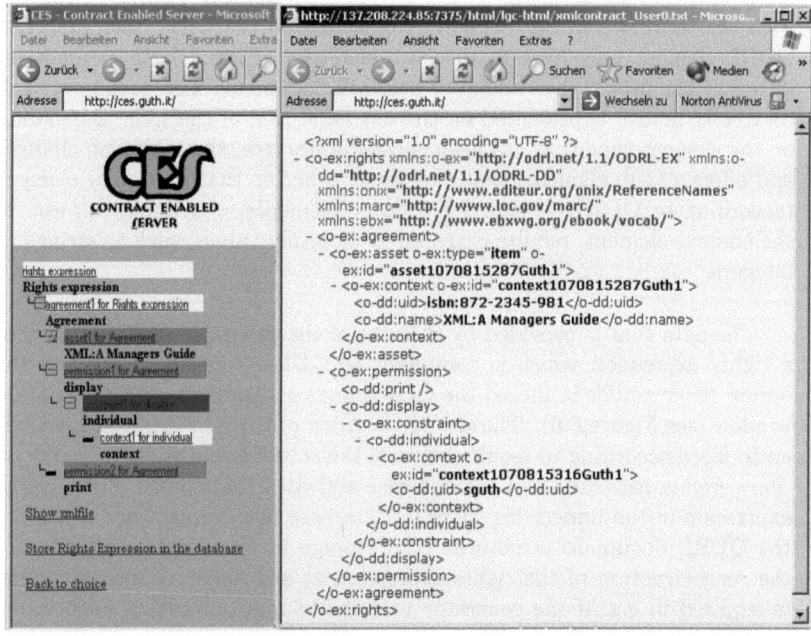

Figure 6.6: Display and store generated ODRL rights expression

The rights expression generator is a very flexible tool in terms of adding new rights expression elements and adapting the graphical user interface. Besides storing rights expressions, the underlying relational database is also used to customise the generator. A large number of properties can be set upon running the tool. To provide application or domain–specific vocabulary for electronic contracts the ODRL data dictionary (see Section 3.4.1) can be extended by additional permission–, constraint–, requirement–, condition– or context elements, i.e. XML tags and attributes. The generator implementation supports the extension by simply adding the new vocabulary into the respective data base table. The help texts to all ODRL elements can also be modified via the data base. Apart from handing an extended ODRL data dictionary the generator supports the integration of already existing namespaces (e.g. Dublin Core [Dub01] or LOM [IEE02]).

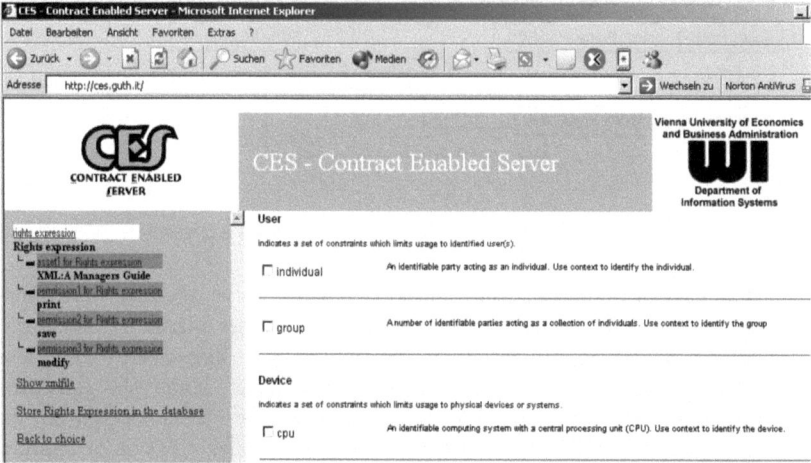

Figure 6.7: Choosing constraints via the customised generator GUI

ODRL is a very flexible language. On the one hand, this means that rights expression composers can be very free and inventive when they formulate rights expressions. On the other hand, the semantics of large and multiply nested rights expressions is difficult to interpret. Therefore, the generator allows to restrict the number and type of subelements which can be customised via the underlying data base. Finally, the generator provides an interface to resource and user management, and thus can be easily integrated into an existing platform.

6.2.2 Class Hierarchy

The rights expression generator reuses ActiWeb (see Section 6.1.2), a library of components for Internet applications written in XOTcl. Among others, the component xoComm of ActiWeb comprises the classes Httpd and Place::HttpdWrk which provide HTTP server functionality. For the rights expression generator the two subclasses ODRLHttpd and ODRLHttpd::Wrk have been developed. They implement an extended HTTP server with the functionality of the rights expression generator (see Figure 6.8). The ODRLSQLWorker handles all data base transaction. In case a different data base or storage mechanism shall be used, only this class and its methods have to be adapted. The distinction between the generator application and the data base access is a design decision that shall provide independence from a specific data base product or mechanism.

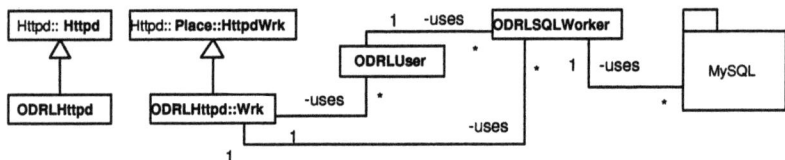

Figure 6.8: Reused software packages in the rights expression generator

The classes ODRLUser and ODRLHttp::Wrk receive all data base functionalities of the ODRLSQLWorker class via *per-class mixin*[10]. The class ODRLUser is additionally used by the ODRLHttpd::Wrk class. The class ODRLUser is an abstraction of the user management. Thus, the generator is able to show a personalised interface to each user. This is required, for example, to show rights holders *their* resources for offer creation or to show the users *their* history of created rights expressions, and finally to actually allow the user to work with the generator.

The classes in Figure 6.9 represent all ODRL tags that are supported by the generator. Each tag class receives common functions from the classes ODRLStack, ODRLSafety, and ODRLTag. The ODRLStack and ODRLStack

---

[10]The per-class mixin is an object-oriented concept provided by XOTcl that enables dynamically adding methods and instance variables of a second class to a first class at runtime. Here, the class ODRLUser dynamically receives all functionalities of the class ODRLSQLWorker. To read more about per-class and per-object mixins, please refer to [NZ99a, NZ99b]

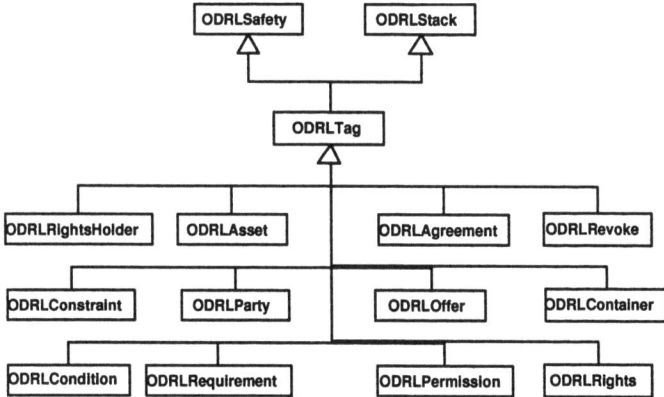

Figure 6.9: Class hierarchy of ODRL specific elements

classes provide the methods push() and pop() for the handling of objects in a stack. The class ODRLTag is mainly responsible for building the resulting XML document, i.e. the ODRL instance. It provides the functionality to create new elements according to the ODRL specification, as well as to delete elements from the XML structure. Each subclass of ODRLTag, such as ODRLPermission, ODRLParty, etc., has tag–specific methods and instance variables for e.g. the creation or HTML representation of that specific tag type.

6.3 The Rights Expression Interpreter

This section presents the implementation of a rights expression interpreter component. The rights expression interpreter has been developed for this doctoral thesis and is called xoREL (from: rights expression language interpreter component coded in XOTcl). This thesis has drawn special attention to the development of the rights expression interpreter. This component provides the interpretation of rights expressions, which is part of the *decoding* stage in the rights expression communication model (see Section 5.1.2). xoREL implements the concept of the application–specific contract schema (see Section 4.6). The interface to the contract schema is provided by the CoSa API (see Sections 4.6 and 10.1) that has been fully implemented by xoREL. The CoSa API ensures an easy access to rights expressions and its processing in software services.

6.3.1 Functional Description

xoREL interprets REL instances for further processing. In other words, rights expressions are transformed into the application–specific contract schema (CoSa) developed in Section 4.5 which supports the usage scenarios *access control* and *accounting*. I chose the usage scenario access control, because in DRM systems rights expressions are processed mainly in access control services, as exemplified in Section 2.3.3. There, rights expressions (licenses) are delivered together with the content to the customer who desires to access the content. A secure viewer is able to interpret the license and interact with the access control mechanism and regulates access to the content. The case study presented in Chapter 7 shows such processing of rights expressions in an access control mechanism.

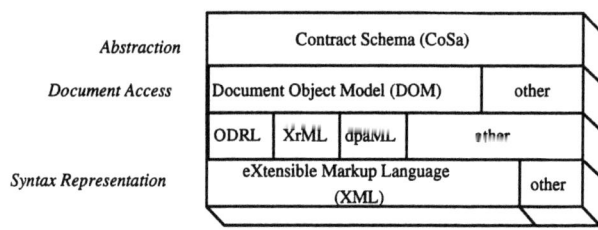

Figure 6.10: Functional layers of XML–based rights expressions

The generic contract schema introduced in Section 4.6 is an abstract, implementation–independent technology. In Figure 6.10 the generic contract schema is related to current technology (respectively standards) for expressing and processing rights. Very often XML is used for the syntax representation of rights expressions (see Section 3.2) and therefore the XML standard builds the basis of the hierarchy. Alternatively, other frameworks for syntax representation can be used. For the definition of certain XML document types either a DTD [BPSMM00] or an XML schema [TBMM01, BM01] can be used. ODRL, XrML respectively MPEG 21 REL are defined in XML schema documents. Other present or future formats for rights expressions can be used instead of the named RELs and are thus not excluded from the CoSa concept. In our implementation the document object model (DOM) (see Section 6.1.3) for the document access. However, other technologies can be used as well, e.g. XPath [CD99]. The CoSa provides an abstraction layer for various representations, document types, and access technologies for rights expressions. Each representation of rights expressions can be mapped to CoSa if the respective interpreter is available. CoSa enables a consistent processing of rights information. The mapping from various formats to one generic representation provides a "contract interface" that supports openness and interoperability. Applications that process rights expressions are independent of the underlying rights expression representation.

A correct transformation of rights expressions requires the syntax and semantics adoption of the respective rights expression language and the contract schema that the REL instances shall be mapped to. For each rights expression language and each application–specific contract schema the according implementation work has to be provided. Details on the implementation concept can be found later in this section. The specifications and requirements of rights expression languages can have peculiarities that have to be considered, for example, which mandatory elements exist in the respective language or what the semantics of a certain element constellation are. ODRL, for example, provides multiple ways to express one single rights expression. Therefore, an interpreter of this language has to consider these facts during the interpretation. Rights expression languages should avoid ambiguity and inconsistency in REL instances. Since there are no formal semantics for existing rights expression languages, there is definitely no completely consistent and unambiguous REL today. When implementing xoREL, several interpretation details that have not been clearly stated in the REL specification had to be discussed with the REL specification authors.

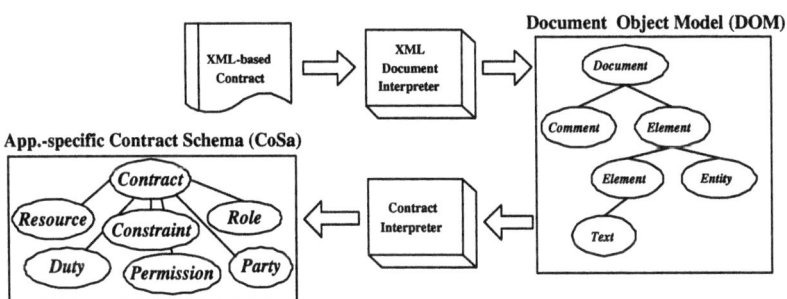

Figure 6.11: The interpretation process

Figure 6.11 depicts the interpretation process of XML–based rights expressions. First, an XML document interpreter (XML parser) is required that reads out the elements of an XML document. For the XML parser it is not relevant that the document semantically represents a contract and simply identifies all XML elements. To build a DOM–tree the XML parser at least needs to implement DOM level 1 (see Section 6.1.3). The contract interpretation is layered on top of the resulting DOM tree. In this step the raw XML document elements are assigned to contract semantics. The contract interpreter maps the elements from the XML document according to the language specifications to the contract schema. This logic has to be programmed for each rights expression language and for the respective contract schema. The respective contract schema (here, application–specific CoSa) then represents all contract information as objects in a flat contract tree, i.e. a tree that has one root object and all other contract objects nested below. The root object is the *Contract* object that "aggregates" all other contract objects, such as *Party, Resource, Permission, Constraint, Duty,* and *Role* on the next lower level. The contract schema can now be queried (by the CoSa API) and further processed in software services. The CoSa API is independent of the contract schema, i.e. in case objects are added or deleted from the contract schema, the same API is used to query the contract information. In Section 7 the processing of contract information in various applications is exemplified.

6.3.2 xoREL Packages and Classes

xoREL is an implementation of the rights expression interpreter component which supports the interpretation of ODRL instances. ODRL has been chosen for this implementation because ODRL is an open source, freely available product and that has no licensing requirements. Furthermore, ODRL has a straight forward simple approach to the expression of rights and is an accepted REL in the research community.

The xoREL implementation consists of two XOTcl software packages: the `reInterpreter` package and the `contract` package. The `contract` package (see Figure 6.12) comprises all CoSa object types. Instances of these object types store the information of the interpreted rights expressions. All object types inherit instance attributes and methods from the abstract class `CoSaObject`. The methods `stdMsg(..)` and `print(..)` facilitate a logging functionality respectively the printing of an object with its name, attributes and attribute values. The attribute `relations` stores all relations of the respective object in a list of ⟨type cosaObject⟩–pairs (see Section 4.6). Table 6.1 shows the possible relations of all CoSa objects.

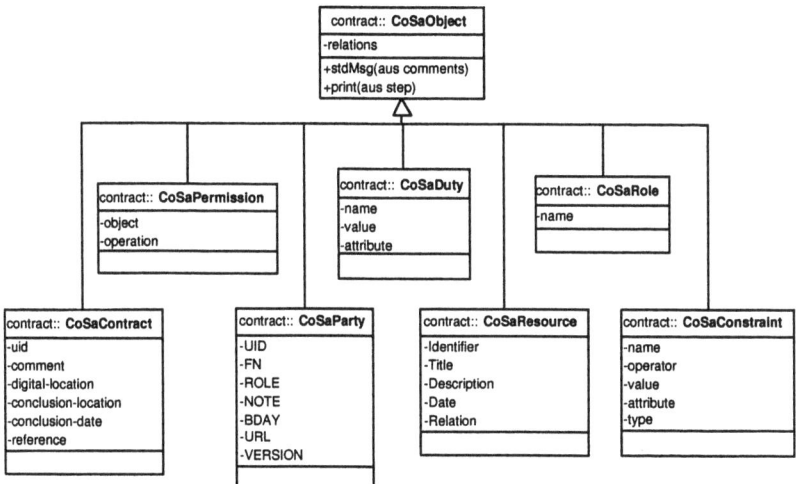

Figure 6.12: Classes of the package contract

These relations have been derived from the application–specific contract schema (respectively the data model developed in Section 4.5.2).

A CoSa contract at runtime comprises various related instances from subclasses of CoSaObject, such as CoSaContract, CoSaParty, CoSaResource, CoSaPermission, CoSaDuty, CoSaConstraint and CoSaRole. Each of these classes provides definitions for its individual attributes. The attributes of the classes CoSaResource and CoSaParty are determined by the standards Doublin Core [Dub01] respectively vCard [HF98] (this is the reason for the different spelling of the attributes). The mapping of ODRL elements to these attributes is shown in Section 6.3.3.

The package reInterpreter defines the CoSa API interface in the class RELContract. Every subclass of RELContract, e.g. ODRLContract or XrMLContract (see Figure 6.13), has to implement the abstract methods of the CoSa API. Please find more information on the CoSa API in Section 4.6, and a detailed description of all methods in Appendix B (Chapter 10.1).

From	To	Relation	Role (from)	Role (to)
Party	Resource	contr. permission	control_perm	controlled_by
Party	Permission	is assigned to	has_perm	granted_to
Party	Permission	grants	grants	granted_by
Party	Duty	is assigned to	has_duty	duty_of
Party	Role	is assigned to	has_role	role_of
Permission	Resource	refers to	refers_to	rel_perm
Permission	Role	is assigned to	perm_of	of_perm
Permission	Constraint	constraints	has_constr	constr_of
Role	Constraint	constraints	constr_by	rel_constr
Duty	Constraint	constraints	with_constr	of_duty
Contract	Resource	comprises	agg_child	agg_parent
Contract	Party	comprises	agg_child	agg_parent
Contract	Permission	comprises	agg_child	agg_parent
Contract	Duty	comprises	agg_child	agg_parent
Contract	Role	comprises	agg_child	agg_parent
Contract	Constraint	comprises	agg_child	agg_parent

Table 6.1: Possible role names in application–specific CoSa

In this thesis the ODRLContract class has been implemented that provides an ODRL interpreter conforming to the ODRL version 1.1 [Ian02b] and that implements the CoSa API. An instance of ODRLContract, e.g. op can receive extra functionality via per–object mixins. Here the instance op receives the functionality of the classes vCard, DublinCore, and ACCoSa that are also part of the reInterpreter package. The mixin of the two classes vCard and DublinCore determines the mapping of ODRL elements to vCard respectively DublinCore attributes in all CoSaParty respectively CoSaResource instances. This means that the instance variables of CoSa objects can be assigned dynamically and also other metadata standards e.g. LOM could be used instead. XOTcl allows this dynamic assignment. Therefore, the instance variables of the classes CoSaParty and CoSaResource (shown in Figure 6.12) can change their names upon mixins. Furthermore, the mixin of the class ASCoSa determines the application–specific contract schema that the contract data shall be mapped to. In this implementation the application–specific CoSa (respectively the class ASCoSa) handles the usage scenarios accounting and access control. For a different application–specific contract schema another class such as the ASCoSa has to be implemented.

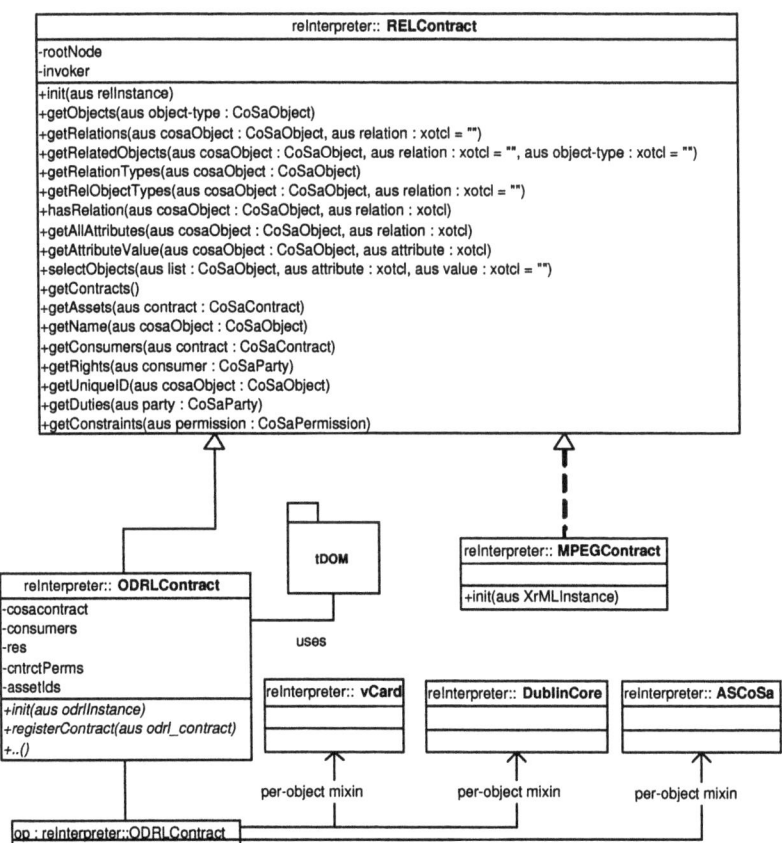

Figure 6.13: Classes of the package reInterpreter

To interpret ODRL documents, xoREL uses tDOM [LA03]. The class ODRLContract assumes that the rights expression to be interpreted is valid, i.e. has successfully been unwrapped. An interpreter of another REL (e.g. XrML respectively MPEG 21) can be added by writing a subclass of RELContract that maps the respective REL instances to the contract schema.

6.3.3 Mapping ODRL Elements to the Contract Schema

In this section a detailed description of the mapping from the ODRL schema to the contract schema is given. It also names the ODRL tags that are currently not mapped onto the contract schema. xoREL currently handles the below listed elements of the ODRL foundation model (see Appendix A, Chapter 9). In the following, the ODRL elements are commented with respect to their mapping onto the interpreter's contract schema:

- *Context.* The ODRL context element provides a large amount of contract data. The context element describes ODRL parties, ODRL asset, ODRL agreements, and ODRL offers in more detail. Depending on which entity the contract element is assigned to, the context information is handled differently. The context element is not represented by an individual CoSa object, but rather fills the instance variables of a large number of other CoSa objects. Please refer to Asset, Party, and Agreement/Offer for the exact mapping.

- *Asset.* The context information of assets is mapped to the Dublin Core [Dub01] vocabulary which then denotes the instance variables of CoSaResource objects. For details, please confer to Table 6.2. The comments in the table describe the Dublin Core semantics of the respective element, and names possible values.

- *Party.* The context information of assets with the exception of *role* is mapped to vCard[HF98] vocabulary which then denotes the instance variables of CoSaParty objects. For role a CoSaRole object is created; the value of role is written into the instance variable *name* of this object and the CoSaRole object is related (via **relations**) to the respective party. For details, please confer to Table 6.3. The comments in this table describe the vCard semantics of the respective element, and name possible values.

- *Rights.* The rights entity is not represented as a distinguished object. The rights information can be found in CoSaPermission objects.

- *Agreement/Offer.* The context information of agreements, and offers is mapped onto language–neutral vocabulary which form the instance variables of CoSaContract objects. For details, please confer to Table 6.4. The comments in this table describe the semantics of the respective element, and name possible values.

- *Permission.* The ODRL term 'permission' is mapped to the 'operation' instance variable of CoSaPermission objects. A permission in CoSa comprises the two instance variables operation and object. A permission is always an ⟨*operation* − *object*⟩ pair. The ODRL permission term is mapped onto language–neutral vocabulary. Unfortunately, no standard exists for the contract description.

- *Rights Holder.* The rights holder element is mapped to the instance variable *ROLE* (contract role) of CoSaParty objects. It determines the role of the ODRL party in the current ODRL instance (consumer or rights holder). Note that ODRL does not provide the possibility to nominate a beneficiary who is neither consumer nor rights holder.

- *Constraint/Condition.* ODRL distinguishes between constraints and conditions. Conditions are the opposite of constraints. Conditions specify exceptions that, if they become true, expire the Permissions (see also 3.4.1). These constructs, however, are not supported by other rights expression languages, e.g. XrML [Con00]. Therefore, in CoSa, ODRL constraints and ODRL conditions are mapped onto instances of the generic CoSaConstraint class that keeps the information about whether it was an ODRL condition or constraint. The constraint type has the instance variables *name, operator, value, attribute* and *type* that facilitate to store both conditions and constraints. An example object would comprise the following values: *name* = datetime, *operator* = ">", *value* = "2004-12-31T00:00:00", and *type* = constraint (empty *attribute*). ODRL does not explicitly provide operators. Therefore, an expression such as the following has to be matched to the generalised attributes shown above by the interpreter.

    ```
    <datetime>
        <end>2004-12-31T00:00:00</end>
    </datetime>
    ```

- *Requirement.* ODRL requirements are mapped to CoSaDuty objects. The CoSaDuty class has the instance variables *name, value* and *attribute* that facilitate the storage of one duty, such as *name* = prepay, *value* = 200.0, *attribute* = €. Note that in the application–specific CoSa, duties can be related to constraints, such as "payment until 31.12.2004", but the ODRL language has not designated conditions for duties.

ODRL	CoSaResource	Comment
uid	Identifier	Formal identification systems include but are not limited to the URI, URL, DOI, and ISBN.
name	Title	A name given to the resource.
remark	Description	An account of the content of the resource, such as an abstract, table of contents, etc.
date	Date	A date of an event in the life cycle of the resource, such as creation date.
reference	Relation	A reference to a related resource (e.g. a string of a formal identification system.)

Table 6.2: Mapping of ODRL asset context to CoSaResource objects

The occurrences in the document, i.e. the interrelations between the ODRL entities is represented by the relation–entries in the **relations** attribute of the CoSa objects. The ODRL sequencing and ODRL inheritance functionalities are not provided in the current implementation. Whereas ODRL inheritance does not supply additional expressiveness, the ODRL sequencing mechanism can be useful and will be addressed in future versions of xoREL.

In the current implementation, all context elements that are not mentioned in the mappings are stored with the ODRL term as instance variables of the respective CoSa object. The CoSa API allows to query these variables and their values, but their semantics is not defined. This makes clear that a rights language would improve its semantics by reusing existing description standards for their entities, such as Dublin Core, vCard, etc., instead of defining own context elements. In ODRL the context elements have different and unclear meanings, depending on which entities they are assigned to. The two ODRL elements Digital Signature and Encryption Digest/Key are not yet mapped to CoSa objects. However, the interpreter "finds" the elements. To interpret and process them, only the respective code extensions and interfaces have to be provided.

ODRL	CoSaParty	Comment
uid	UID	A value that represents a globally (or in the closed domain) unique identifier corresponding to the individual, such as x509 certification serial number.
name	FN	A formatted text corresponding to the name of the object the CoSa object represents, such as Steffi Graf.
remark	NOTE	To specify supplemental textual information or a comment that is associated with the CoSa object.
role	ROLE	To specify the contract role of the individual, consumer or rights holder.
date	BDAY	Specifies the birthdate of a party. Please note that the date in an ODRL context element is not necessarily describing the birthday of that party.
dLocation	URL	To specify a uniform resource locator associated with the individual.
version	VERSION	To specify the version of the vCard specification used to format this CoSaParty object.

Table 6.3: Mapping of ODRL party context to CoSaParty objects

6.4 The Rights Expression Wrapper and Unwrapper

Rights expressions have to be exchanged securely between partners. Depending on the practical applications various security services have to be applied to the rights expression, such as integrity, authentication, or digital signature verification. These services are provided by the rights expression wrapper and unwrapper. The rights expression unwrapper is the complementary component to the rights expression wrapper. It unwraps the rights expression does the extrinsic checking of the digital contract, such as check-

ODRL	CoSaContract	Comment
uid	uid	A value that represents a globally (or in the closed domain) unique identifier corresponding to the contract.
remark	comment	A supplemental textual information or a comment that is associated with the contract.
date	conclusion-date	Please note that the date in a context element is not necessarily describing the birthday of that party.
dLocation	digital-location	The digital location where the contract can be found, e.g. an URI, URL, etc.
pLocation	conclusion-location	The physical location the digital contract was concluded, e.g. name of the city.
reference	reference	A link to additional information about the contract, such as legal information.

Table 6.4: Mapping of ODRL agreement/offer context to CoSaContract objects

ing the contract integrity or the authentication of the rights expression sender or verifying the digital signature of the rights expression (see Section 5.2).

6.4.1 Functional Description

In the current implementation the wrapper facilitates to digitally sign a rights expression and the unwrapper facilitates to verify the digital signature of the wrapper. The wrapper and unwrapper reuse the open source openssl implementation of the *OpenSSL Project*[11] that provides the cryptographic functions required for the digital signature.

---

[11] See: http://www.openssl.org/

6.4.2 Class Hierarchy and API

The two components, rights expression wrapper and rights expression unwrapper, are implemented as XOTcl software packages reWrapper and reUnwrapper (see Figure 6.14). In the actual implementation the wrapper simply includes the class Wrapper which provides among others the methods init(..) and wrap(..). When creating a new Wrapper instance, the method init(..) is called with one parameter. The parameter is the rights expression that shall be wrapped. Calling the method wrap(..) invokes a number of other methods that are necessary for wrapping the rights expression. In the current implementation, first a hash is created from the rights expression, then the hash is signed and subsequently rights expression and signature are packed in an archive. The actions can also be performed independently of each other by calling the method names hash(..), sign(..) and pack(..) that are also defined in the class Wrapper.

The unwrapper in return, performs all actions that are necessary to unwrap the packaged rights expression. The only class Unwrapper in this package, includes among others the methods init(..) and unwrap(..). When creating a new Unwrapper instance, the method init(..) is called with one parameter. The parameter is the archive that shall be unwrapped. Calling the method unwrap(..) invokes the unpacking of the rights expression and license from the archive, verifies the signature of the hash (i.e. is the rights expression signed by a trusted party, here the wrapper), crates a second hash from the rights expression, and finally compares the unpacked hash with the newly created one. Again, all methods that perform the various checks can also be called independently of each other by the names unpack(..), verifySignature(..), hash(..), checkIntegrity(..). For a full description of this short wrapper interface, please refer to Appendix B.(Section 10.3). The third package Utils does the binding to the OpenSSL implementation and provides the wrappers with utilities that use OpenSSL. As both wrapper components need to access these utilities, they are arranged within a separate package.

The wrapper and unwrapper interface can be extended by adding methods of the Wrapper respectively Unwrapper classes. The methods in these classes depend on the application fields of the rights expression exchange framework. An actual framework implementation needs to support all functionalities to check the contract (respectively rights expression) validity (see Section 4.8). Accordingly, potential additional interface methods of the

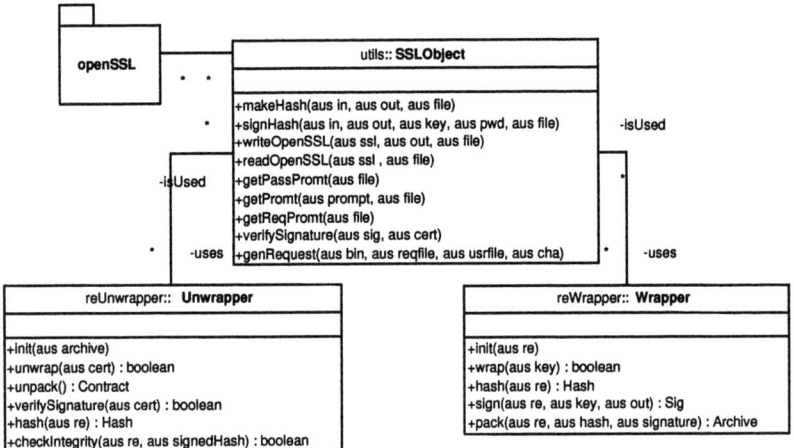

Figure 6.14: Packages with wrapping respectively unwrapping functionality

wrapper are `encrypt(..)`, `checkRELofEnoding(..)`, `checkResourceAvailability(..)`, etc.

6.5 The Mediator

As described in Section 5.2.1 the mediator is the "glue code" between the framework components (or packages) and reused software, such as access control mechanisms. For example, if a platform shall be prepared to unwrap contracts, interpret them, and implement their content to an access control mechanism, the mediator has to combine the functionality of the unwrapper, the interpreter, and the access control mechanism. Technically, here the mediator is an XOTcl class that imports other packages (via `package require`) respectively classes, such as the `Unwrapper` class of the `reUnwrapper` package, the `ODRLContract` class of the `reInterpreter` package, and the `RightsManager` class of the xoRBAC[12] package (see Figure 6.15). The `Mediator` class initiates instances of all three services and uses their API to unwrap the received contract (respectively rights expression), to interpret it, and to transform the contract information into access control information. Therefore, the execution of the `Mediator` class also determines

---

[12]xoRBAC is a role–based access control (RBAC) mechanism. It is introduced in detail in Section 7.1.

the workflow of the usage scenario. The XOTcl code of such mediator is shown in Section 7.2.4. A second mediator is necessary is a platform requires to formulate and digitally sign rights expressions. In that case the mediator has to intercede e.g. between the generator, the web server, a contract database and the rights expression wrapper.

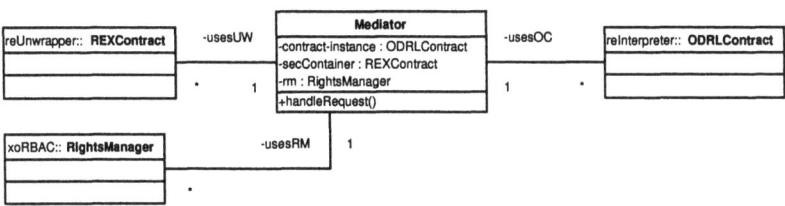

Figure 6.15: The mediator, using framework components and other packages

Each usage scenario requires an individual implementation of the mediator, i.e. the mediator implementation is application–specific. The generator implemented in this work, for example, does not provide an API because it is a web application. Still the wrapper functionality has to be integrated. In this case the web application is extended by another ODRLHttpd::Wrk method (see class diagram in Figure 6.8) which handles the wrapping (signing) of licenses (see Chapter 7). This method represents the mediator. Consequently, the framework implementation can not provide a generic mediator implementation, as the integration of components is individual. Still, the framework includes a mediator template as well as mediator sample implementations.

6.6 Implementation Assumptions

The implementation of the rights expression exchange framework assumes, that the software that uses the framework components absorbs the responsibility for ids. This includes mapping resource ids and user ids to the respective objects and checking resource ids and user ids in terms of correct spelling, etc. For example, the generator requires user and resource ids to filter all offers and authorizations of a certain user respectively show to a content provider *his/her* resources. Furthermore, the implementation implies, that a public key infrastructure, and a reliable certificate handling is given. This includes, that e.g. the wrapper signs the rights expressions with a key, that is accepted by the partner DRM platforms. If the users do not have certificates from accepted certification authorities, a certification authority has to be installed and operated (e.g. to issue consumer certificates). The implementation does not require a central server. Each computer that runs the rights expression exchange framework is a technically equal peer in a network of (cooperating DRM) systems.

When evaluating access requests, the implementation currently is able to handle time constraints and constraints that refer to the IP address of a network computer, i.e. the implementation facilitate to check environment attributes. The checking of context constraints in our application is handled by the access control mechanism xoRBAC that provides sensors to receive the current time and IP address (see Section 7.1). To check constraints that require information such as, *"How often user X has used resource Y?"*, the framework needs to be integrated into a DRM system. For the integration either a mediator has to be implemented that queries the DRM database or the access control mechanism is extended with sensors for the DRM database.

The implementation facilitates rights expression exchange in the usage scenarios access control and accounting, and assumes that the rights expressions are formulated in ODRL [Ian02b]. If other usage scenarios or other rights expression languages shall be supported, the code has to be extended. Whereas the extensions for a different application–specific CoSa or for the adoption of a different metadata standards are not very complex, the implementation of an interpreter for a different rights expression language (e.g. MPEG REL) is fairly costly. For a technical check list please also refer to Section 5.2.2.

6.7 Related Work

This chapter introduces other implementations of rights expression interpreters and contract management systems. The closest related work is the rights expression interpreter XrML SDK of today's second important rights expression language XrML.

XrML SDK

The eXtensible rights markup language (XrML) [Con00] is a rights expression language developed by ContentGuard [13]. XrML Version 2.0 was selected by the moving picture experts group (MPEG) as the basis for development of the MPEG 21 Part 5[14] standard. ContentGuard has released a XrML Software Development Kit (SDK) that includes an example license interpreter and an example condition validator.

- The *license interpreter* basically provides a single method: *validateGoal(PrincipalList, RightsList, ResourceList)*. This method fetches all conditions from an XrML document that are linked to the triple ⟨principal, resource, right⟩. For example, if the respective XrML license contains the triple ⟨Mary, hit.mp3, play⟩ the method fetches all conditions associated with this triple (e.g. a restriction to play the music file for five times only). All conditions extracted via the validateGoal method are further processed by the *condition validator*.

- The *condition validator* checks if the corresponding conditions are met, in the example mentioned above, it decides whether Mary may play the hit.mp3 or not. For the time being, the XrML condition validator is capable of checking two constraints: time interval and exercise limit.

The XrML license interpreter and the condition validator are closely coupled and are not designed to operate separately. The interpreter does not provide additional functions to make contract information available to other applications, such as access control services. Thus, the current version of the XrML SDK is a proof–of–concept implementation that is focused on one particular application of the rights expression language XrML. In contrast to XrML SDK, in this thesis a general approach for the interpretation of rights expression languages has been developed. The CoSa can be applied to extend arbitrary applications (providing C or Tcl linkage) with contract processing abilities. Thus, the CoSa itself is independent of certain rights expression languages and from the applications that use/process contract data.

---

[13]See: http://www.contentguard.com/
[14]The ISO/IEC working group in charge of the development of standards for coded representation of digital audio and video, http://www.chiariglione.org/mpeg/)

ODRL Implementation

Renato Iannella, the founder of the ODRL initiative, implemented at IPR Systems[15] an application–specific ODRL generator and ODRL interpreter for the Colis [16] project [Ian03c]. Not all elements, but a smaller profile of ODRL [Ian02a] is supported by the implementation that allows users to enter offers respectively allows the consumers to select content and generate agreements [Ian03a]. According to the contract, an ODRL license and the content are packaged together and downloaded by the user. In the COLIS project, when the content is accessed by a user, a component (including interpreter and access control mechanism) checks if a certain role is assigned to this user before allowing access. Consequently, the Colis implementation is able to handle group constraints. Unlike as in this thesis, the documentation of the implementation does provide a generic approach for the processing of rights expressions. Furthermore, in this thesis all ODRL elements are supported for the generation of rights expressions.

Others

Park and Sandhu propose a high–level model for the definition of usage control policies [PS02b]. Usage control (UCON) works on the principle that digital objects are encapsulated in a secure "digital container". Information within such a digital container can only be accessed through specific (tamper–resistant) soft– and/or hardware devices by feeding in a set of access rights approved by the originator of the corresponding container. The set of access rights can be regarded as a license or a contract between the originator and the recipient/consumer. However, their model is, defined on a high level of abstraction and must be refined before it can serve as a basis for the definition of actual UCON policies. In [PS02a], Park and Sandhu describe an approach to combine usage control and originator control. Originator control has already been mentioned in [Lan81]. It is a concept which requires that recipients obtain the originator's approval prior to the re–dissemination of digital objects. In Park and Sandhu's approach, "licenses" are digitally signed certificates defining the usage rights for digital objects. Users can access digital objects only according to their license. Tickets are used to transfer "re–dissemination" rights for digital objects.

Shand and Bacon [SB02] present a contract framework that includes an abstract contract protocol for contract exchange and an accounting language

---

[15]See: http://www.iprsystems.com/
[16]See: http://www.colis.mq.edu.au/

(based on the Python scripting language) for the specification of accounting policies. Contracts define the resources that are exchanged between contracting parties, e.g. CPU time, network bandwidth, or money. Contracts must be signed by all contract parties to be valid. A peculiarity of their approach is that trust is treated as a special type of resource which influences the conclusion of a contract. The trustworthiness of a certain party is continuously adapted according to her/his contractual fidelity.

Chapter 7

Case Study of the Rights Expression Exchange Framework

This chapter presents the application of the rights expression exchange framework that has been designed and implemented in this thesis (see Chapters 5 and Section 6). Normally, not all four components of the framework are located on the same machine. Usually, the rights expression generator and the wrapper are located on the sender's platform and the unwrapper and interpreter are running on the receiver's platform. However, due to the component–based approach of the rights expression exchange framework, there can be scenarios where only the interpreter is used and the other components are replaced by foreign components.

Generally, the framework is used for the exchange of rights expressions which state use or access rights of people to digital goods or services. The exchange of these rights expressions takes place between two or more interoperating (trusted) partners. There are various general scenarios in which rights expressions need to be exchanged, for example:

1. A consumer has concluded a contract with the marketer X of certain services (communication or multimedia). For example, the contract allows to use different service providers for sending SMS, faxes, emails, MMS, ring tones, logos, etc. via the Internet for a certain time, or to access online platforms to play videos, music files, games, etc. within

the next month. Each time the consumer desires to access a service s/he has to present the contract to the service provider. The service providers only accept contracts that have been concluded with certain marketers, e.g. marketer X. The rights expression exchange framework can provide all functionalities that are necessary for this scenario: formulating the contract and signing it is handled by the generator and the wrapper that are running on the marketer's system. Each service provider is runs the unwrapper and interpreter component to verify the marketer's signature on the contract in order to subsequently interpret the consumer's access rights. The access to the communication respectively multimedia services is then either granted or denied, accordingly to the rights expressions in the contract.

2. User A has certain access rights on platform A or system A (e.g. to the Intranet of a company). Platform A starts a cooperation with Platform B (e.g. a company merges with a second one). From this follows that all users of platform A shall receive the same rights on platform B. For all users electronic tickets (see Section 4.3.1) comprising their personal access rights are issued and transmitted to platform B. Platform B interprets the rights and implements them in their own access control mechanism. In this case, it might be necessary to use a expression generator which is different form that implemented in this thesis, e.g. one that generates rights expressions automatically. However, the wrapper, unwrapper, and interpreter from my implementation could be used for the remaining tasks of the rights expression exchange.

3. A person has a certain role in society or in a community, e.g. that of a student or a pensioner. An commonly accepted institution (e.g. a university or national office) certifies these roles to the respective persons in form of an identification (e.g. a student or pensioner id). Displaying this id entitles these people to receive certain access or usage rights, e.g. going to the library for free, or paying a reduced price for theater tickets. The role in the (electronic) id and possibly some constraints (e.g. expiration) are formulated and signed by the rights expression generator respectively the wrapper running at the certification institution. The providers that accept the ids need an unwrapper and interpreter to verify authenticity of the id and to extract the role from the id. Subsequently, each provider can grant different access or usage rights to the id holders. Note that the extrinsic format of such an id can be an x509 certificate and the rights expression can be stored in the extension field of the certificate.

The above examples display some basic shapes of rights expression exchange scenarios. The rights expression exchange framework supports the drawbacks of today's electronic commerce systems listed in Section 1.1, e.g. the standardised representation of rights expressions, more precisely of contracts, the usage of one contract at various platforms, the processing of rights expressions in various software services, the expression and enforcement of new and rich usage variants for electronic goods, etc. Rights expressions can be processed in various usage scenarios (see Section 4.4.2). As the particular focus of this work is on the usage scenario access control, Section 7.1 gives a short introduction to the processing of access requests by an access control mechanism. The section also addresses the handling of context constraints for usage rights (e.g. play video *five times*) that are frequently used in *new* pricing models for electronic goods. Section 7.2, describes the required rights expression exchange components and their interaction in the technical realisation of the usage scenario access control based on electronic tickets, i.e. the scenario 2 described above.

7.1 Access Control with Context Constraints

A typical example of a rights expression that occurs in DRM systems is "Person X may play the video Y five times on his/her own computer (IP address 137.224.208.84) until the end of this year." To enforce such a rights expression, the access control mechanism has to be able to handle context constraints (in the following called *constraints*). Constraints narrow access permissions, for example by time, location, individual, etc. (please confer also to Sections 4.5 and 3.3.1). In the example the permission *play the video Y* is constrained by location ("his/her own computer"), time ("until the end of this year"), and the number of accesses ("five times").

Handling constraints requires the implementation of constraints and their evaluation at the time of an access request. A constraint is a clause that contains one or more context conditions. A constraint is satisfied iff all its context conditions are met [Str03]. A context condition is a predicate (a boolean function) that consists of an operator and two operands, e.g. $\langle date, <, 12/31/2004 \rangle$. Only if all conditions of a constraint hold, i.e. the overall evaluation of conditions returns "true", the respective permission is granted. Note that the first (or left) operand always represents a certain context attribute (i.e. a property of the environment, such as date), while the second operand may be either a context attribute (e.g. end-date) or a constant value (e.g. 12/31/2004). To evaluate the expression in the previous paragraph the access control mechanism requires the current date, the local computer, and the number of times the video Y has already been played by person X. For this purpose, the access control mechanism uses of *sensors* that return current values for the context condition attributes. For the work at hand, two categories of attributes are be distinguished; environmental attributes and attributes administered by a DRM system database(also see to Section 4.7):

- *DRM System Database.* DRM systems observe all activities that are relevant for the overall rights management in the DRM system database, e.g. how many times a user has accessed a certain resource. Context conditions that define a certain number of usages are very common in DRM applications. To evaluate such conditions the sensors that retrieve the relevant information and DRM system databases have to be available.

- *Environmental Attributes.* Environmental attributes attributes whose values are not administered by the DRM system. These attributes exist independently of DRM systems, rights expressions or contracts. Examples are time, date, IP address, weather conditions, etc. For each required environmental attribute an adequate sensor has to be available to the access control mechanism.

Both categories may include static and dynamic attributes, for example IP address and date (environmental), respectively the user's birth date and the number of access to a certain resource (DRM system knowledge). In section 5.2.1 we have learned that after decoding, the rights expression is available for processing. This section is focused on processing and enforcing the rights expressions in an access control service. Note that the following example is independent of any access control approach. Figure 7.1 depicts the relevant steps that are performed when processing rights expressions in the usage scenario *access control*:

1. *Implement Rights Expression.* In the first step, the rights expressions are retrieved from the rights repository and "fed" to the access control mechanism, i.e. the parties, permissions, and constraints (e.g. person X; play video Y; five times, until the end of the year, on the computer with the IP address 137.224.208.84) are mapped to corresponding policies rules that can be enforced by the access control mechanism. Technically, the permission "play video Y" and the subject X are created.

193

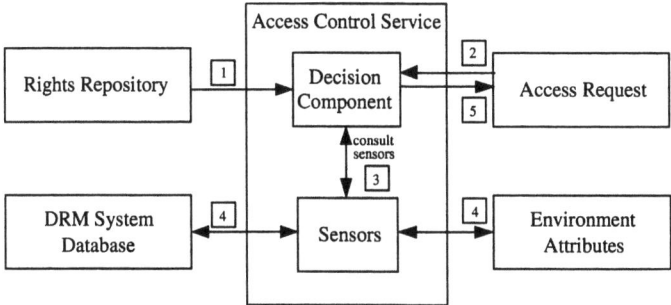

Figure 7.1: Execution of an access request

The permission is assigned to person X. The three constraints (date, IP address, maximum number of accesses) are created and assigned to the permission "play video Y" of person X. The resulting access control policy is depicted in Figure 7.2. This transformation is not performed automatically, but has to be supported by a *mediator* (see Section 6.5).

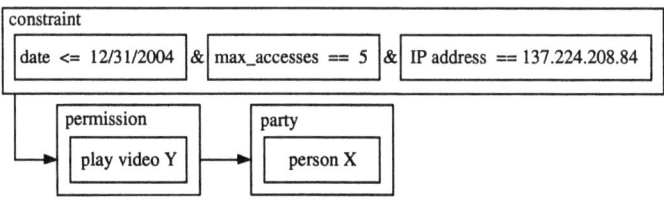

Figure 7.2: Sample access permission with constraints

2. *Access Request.* Person X now aims at executing a certain permission (e.g. play video Y) and triggers the respective access request. The decision component has to evaluate the incoming access request and either grant or deny it. First, the access control mechanism queries the implemented policies for the permission *play video Y* of Person X. If person X does not hold this permission, the decision component denies the request. In the above example the permission *play video Y* is available, but with related conditions.

3. *Sensor Consultation.* To verify time, location, and count conditions, the decision component consults external sensors that are capable of delivering values for the boolean functions (conditions). To ascertain the current time, and the local IP address the decision component consults the environment sensors. For the number of times, that the video Y has already been accessed the DRM system database has to be queried.

4. *Provide State Information.* The respective sensors return the required context information, but do not make any decisions. Let us assume that the current date is 11/11/2004, the IP address of person X's computer is 137.224.208.84, and the video has not been accessed before. *Context functions* [Str03] transform the received values into a readable format for the access control decision component.

5. *Return Evaluation Result.* The decision component receives the context information of the sensors and inserts them into the boolean functions. With the state information received above all boolean functions return true, and the access request can be granted. If one function returns "false", the access to the video Y is denied.

The Access Control Mechanism xoRBAC

xoRBAC [NS01, NS03a] is an access control mechanism that, among other things, supports the handling of context constraints. It is used as access control component in the subsequent framework application example. xoRBAC provides a role–based access control (RBAC) service that can be used on Unix and Windows systems in applications providing C or Tcl linkage. xoRBAC is well–suited to be used within a component framework. While originally developed as an RBAC service, xoRBAC was extended to provide a multi-policy access control system which can enforce RBAC–, as well as DAC– (discretionary access control) or MAC– (mandatory access control) based policies including *conditional permissions*. With respect to this thesis, the *dynamic constraint management* subsystem is of central significance. It comprises the *environment mapping*, which captures context information via sensors, and the *constraint evaluation*, which checks if the collected values match the context constraints associated with a certain conditional permission. Thus, it allows for the definition and enforcement of context constraints.

The subsequent paragraphs, describe features of xoRBAC that are necessary to enforce the access control policies with context constraints as described above. A *context constraint* is defined through the terms context attribute, context function, and context condition:

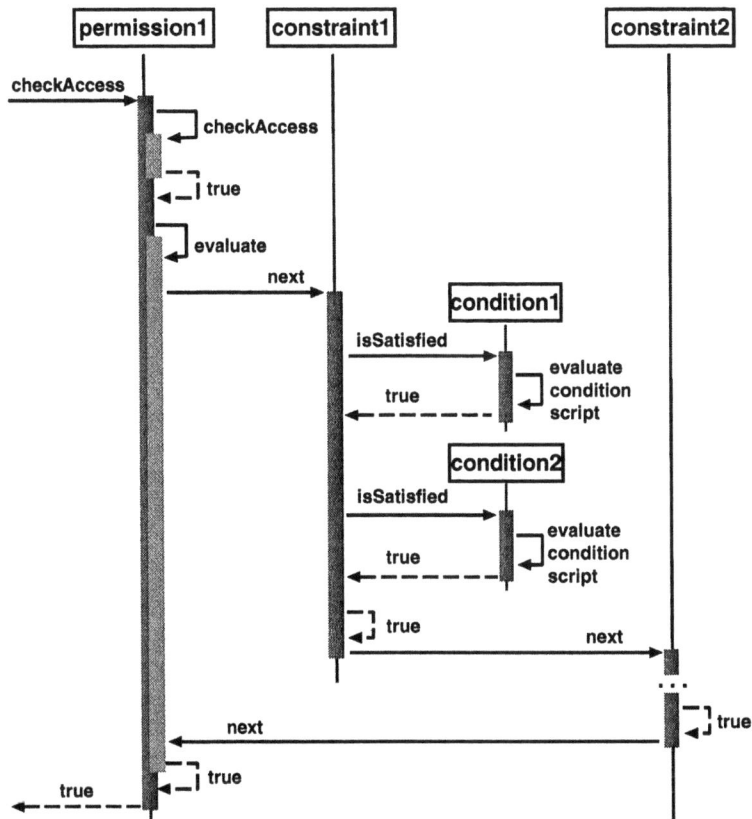

Figure 7.3: xoRBAC access control decisions with context constraints

- A *context attribute* represents a certain property of the environment whose actual value might change dynamically (like time, date, or session-date), or for different instances of the same abstract entity (e.g. location, ownership, birthday, or nationality). Thus, context attributes are a means to make (exogenous) context information explicit.

On the programming level each context attribute *CA* represents a variable that is associated with a $domain_{CA}$ which determines the type and range of values this attribute may take (e.g. date, real, integer, string).

- A *context function* is a mechanism to obtain the current value of a specific context attribute (i.e. to explicitly capture context information). For example, a function *date()* could be defined to return the current date. Of course a context function can also receive one or more input parameters. For example, a function *age(subject)* may take the subject name out of the ⟨subject, operation, object⟩ triple to find out the age of the subject which initiated the current access request, e.g. the age can be acquired from some database.

- A *context condition* is a predicate (a Boolean function) that compares the current value of a context attribute either with a predefined constant or another context attribute of the same domain. The corresponding comparison operator must be an operator that is defined for the respective domain. All variables must be ground before evaluation. Therefore, each context attribute is replaced with a constant value by using the according context function prior to the evaluation of the respective condition. Examples for context conditions are $cond_1 : date() \leq "2003/01/01"$, $cond_2 : date() == birthday(subject)$, or $cond_3 : age(subject) > 21$.

- A *context constraint* is a clause containing one or more context conditions. It is satisfied iff all its context conditions are met. Otherwise it returns false.

With respect to the terms defined above, a *conditional permission* is a permission that is associated with one or more context constraints and grants access only if each corresponding context constraints evaluates to "true". Figure 7.3 shows a message sequence chart for access control decisions in xoRBAC including conditional permissions. For a detailed description of xoRBAC see [NS01, NS03a].

7.2 Access Control Decision Based on Electronic Tickets

The rights expression exchange framework usage exemplified in this section handles the following scenario: Platform A generates and digitally signs a license that grants certain access rights to user *M. Strembeck* on a various platforms within a network, e.g. platform B. Platform A sends out the valid license to M. Strembeck. If Mr Strembeck desires access rights on Platform B he chooses the rights license and presents it to platform B to receive access rights. Platform B unwraps and interprets the license and grants or denies the user's access request accordingly to the permissions in the license." This technical use case support a number of higher level applications, e.g. the example, described as item 1 at the beginning of this chapter, but also the following classical DRM application: A DRM platform issues a license that grants access rights to a resource with certain constraints. The platform delivers the license (with or without the resource) in a secure container (see Section 2.3) to the consumer. The secure viewer receives the license and subsequently handles the access request to the resource (that is either in the container or locally stored) and, according to the rights in the license, either renders the resource or not.

The subsequent sections apply the concepts and implemented components of this doctoral thesis to the DRM scenario described above. Section 7.2.1 develops the application–specific CoSa that comprises all necessary contract objects for the scenario. In Section 7.2.2 the adequate licens is created with the rights expression generator. Subsequently, the license is wrapped (see Section 7.2.3). The last section, Section 7.2.4 describes the unwrapping, interpretation and processing of the license. Here, the unwrapper and interpreter that have been developed in this thesis are deployed as well as the access control mechanism (xoRBAC).

7.2.1 Application–Specific CoSa

For the scenario mentioned above the required objects in the license are *Party, Resource, Permission* and *Constraint*. It is assumed that the DRM system is designed to operate with discretionary access control, as licenses are rather issued to individuals than to roles. Therefore, the element *Role* is not required in this DRM application, nor is the element *Duty*. Therefore, for this DRM application the implemented CoSa is satisfactory, yet only the subset of those shown in Figure 7.4 is used. The *Party* objects basically require a unique id; additional attributes are optional. Similarly, the *Resource* objects require a value in the attribute "Identifier". The *Permission* objects must comprise the operation (e.g. display) and the object (resource) the operation refers to. Each permission can be constrained by one ore more conditions. *Constraint* objects comprise the instance variables name, operator, value, attribute, and type (see also Section 6.3.3).

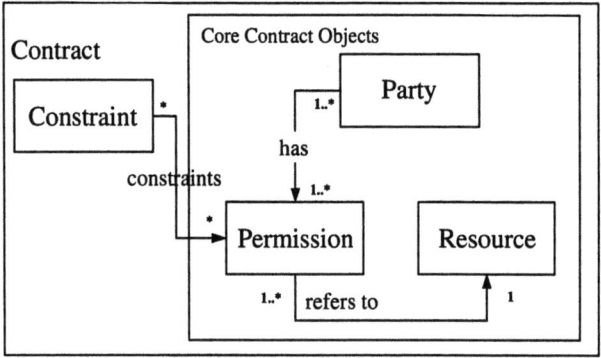

Figure 7.4: The application–specific CoSa

Please note that the rights expression exchange framework is not responsible for the attribute values, i.e. the data in the license. It is assumed that the numbering systems for user ids and resource ids are aligned between platform A and B and that stated access rights, as well as their constraints, are either globally unique or well–understood by platform B. The possible access rights are dependent on the capabilities of the rendering software (e.g. the secure viewer). If the rendering software is able to process the rights "display", "preview", and "copy", then a license comprising the operation "give" will cause an error during rendering. The same is also true for constraints.

7.2.2 Generating DRM–Specific Licenses

According to the developed CoSa, tailored license templates should be provided to the person who fills the license with data (see Section 4.5 and 4.5.3). With the rights expression generator implemented in this thesis, the respective 'empty' templates could be created by the system administrator and stored in the license repository. Users of the generator would restore such licenses and fill them with ids, permissions, and constraints. Figure 7.5 shows, among other rights expressions, a predefined license template for the operation display (license no. 2). Clicking on the modify link would restore the license template and allow value entries by the current user. Storing it again allows to store the copy under a different name (license no. 1).

ID	License short name	Actions
License 1	'display' license for mstrem (ends 2004-12-31)	Sign, Modify
License 2	license-template display	Sign, Modify

List of rights expressions created by user: Guth

Figure 7.5: Provide license templates with the generator

The resulting ODRL rights expression contains a *license*, sometimes also called *digital ticket* (see Section 4.3.1). The license that has been just formulated with the rights expression generator has the following ODRL version 1.1 conform XML code[1]:

```
<?xml version="1.0" encoding="UTF-8" ?>
<o-ex:rights xmlns:o-ex="http://odrl.net/1.1/ODRL-EX"
             xmlns:o-dd="http://odrl.net/1.1/ODRL-DD">
   <o-ex:asset>
      <o-ex:context>
         <o-dd:uid>sguth-9999</o-dd:uid>
         <o-dd:name>Ebook on the framework design</o-dd:name>
      </o-ex:context>
   </o-ex:asset>
   <o-ex:party>
      <o-ex:context>
         <o-dd:uid>mstrem</o-dd:uid>
         <o-dd:name>M. Strembeck</o-dd:name>
      </o-ex:context>
   </o-ex:party>
   <o-ex:permission>
      <o-dd:display>
         <o-ex:constraint>
            <o-dd:datetime>
               <o-dd:end>2004-12-31T00:00:00</o-dd:end>
            </o-dd:datetime>
         </o-ex:constraint>
      </o-dd:display>
   </o-ex:permission>
</o-ex:rights>
```

---

[1] ODRL is defined in two linked XML schemata, the ODRL grammar (prefix: o-ex), and the ODRL data dictionary (prefix: o-dd)

listing grants the rights display to the asset sguth-9999 with the title *Ebook on the framework design* to the party M. Strembeck with the (locally) unique id mstrem. The operation display is narrowed by the following constraint: display may only be executed if the current date one before 12/31/2004.

7.2.3 Wrapping DRM Licenses

Figure 7.5 shows two licenses in the current repository. License no. 1 is a ready–to–use license (see XML serialisation above) that can now be wrapped and transmitted to DRM platform B. In the very right section two actions can be performed with the licenses: *Modify* and *Sign*. The sign action is relevant for the license wrapping. The wrapping functionality is performed by the rights expression exchange framework package reWrapper described in Section 6.4 and is here provided to the user. To integrate the wrapper functionality into the rights expression generator, the reWrapper package has been included into the generator package. The mediator code in this case is written in the extra method wrap(args) of the ODRLHttpd:Wrk that uses an instance of the class Wrapper. The method restores the XML code of the respective license from the repository, signs it with the private key of platform A, packs the license into an archive, and (currently) stores it in a folder, which the creator of the license has access to. Please note that the method wrap(..) serves as a Facade [GHVJ94] for the rights expression wrapper, which subsequently calls all wrapping functions that are adequate for this application. The program in Figure 7.6 below shows extracts from the mediator code within the generator package. The license creator on Platform A sends the wrapped license to M. Strembeck via a secure connection, i.e. HTTP over secure socket layer (SSL) [FKK96].

7.2.4 Unwrapping, Interpreting and Processing DRM Licenses

Let us assume that some day M. Strembeck desires to access the resource "Ebook on the framework design" that is stored on platform B. Therefore, he authenticates himself to platform B with user id "mstrem" and password via *HTTP Basic Authentication* [FGM+99, FHBH+99]. Then he uses an HTML form to specify the resource he desires to access (e.g. display) and to upload his license. Again, the the HTTP connection is secured by SSL. Now, the license has to be processed and the access request has to be evaluated, i.e. either granted or denied. According to the rights expression communication model developed in Section 5.1.2, the processing of the license proceeds with

```
package provide rex::reGenerator
package require rex::reWrapper

package require xotcl::actiweb::webDocument
package require xotcl::actiweb::htmlPlace

Class ODRLHttpd -superclass Httpd
Class ODRLHttpd::Wrk -superclass Place::HttpdWrk

ODRLHttpd::Wrk instproc wrapLicense id {
    set folder "licenses"
    my instvar sign-key
    my instvar statement
    set statement „select Code from odrl.ODRLLicenses where UID ='$id';"
    set result [my dbaccess]
    set channel [open $folder/odrl-license-$id.xml w+]
    puts $result
    close $channel
    Wrapper w $folder/odrl-license-$id.xml
    w wrap $sign-key
}
```

Figure 7.6: Mediator code combining generator and wrapper functionality

unwrapping and interpreting the license. All involved activities from the rights expression exchange framework and environmental components are sketched by the activity diagram in Figure 7.7. The different actors are:

- The *beneficiary*, (or a corresponding client program) who requests a service respectively the access to digital goods and presents a digital license.

- In this case study the *secure viewer* is the *mediator* (see Section 6.5) that coordinates and controls the beneficiary interaction, the rights expression unwrapper and interpreter, as well as the access control service.

- The rights expression *unwrapper* (see Section 6.4) that performs encoding and validity checks on the license.

- The rights expression *interpreter* (see Section 6.3) that parses and interprets the license, and builds a runtime model of the application-specific CoSa.

- The *access control service* which decides if the beneficiary may perform the requested operations according to the permissions granted through the presented license.

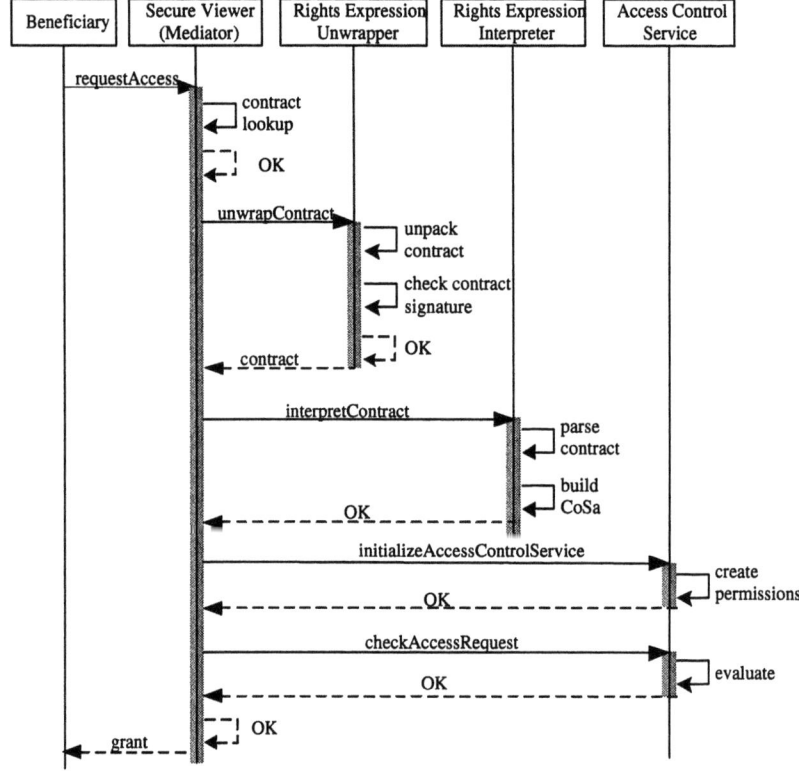

Figure 7.7: Sequence diagram with basic activities of the secure viewer

The contract processing procedure is triggered by the access request of the beneficiary via the HTML form. An access request expresses the demand to perform a specific operation/action on a particular object/resource. In this case study approach, each access request consists of a method call with four parameters ⟨subject, operation, object, contract-location⟩ where the subject is mstrem, the operation is display, the object (requested resource) is sguth-9999 and the contract-location states concrete location of the license that has been uploaded by the beneficiary. The license is fetched by the contract-id via the lookup procedure of the secure viewer. After the contract has been loaded the *unwrapper* component performs the following activities:

- *Unpack License*: unpacks the license from the archive.

- *Check Signature*: verifies the digital signatures of the corresponding license issuer, here platform A, to ensure its integrity and authenticity. The public key of platform A and the information of the applied hash algorithm is required for the signature check. In this case study the public key of platform A is the default key for signature verifications, and the algorithm SHA1 [EJ01] is used as default hash algorithm by the wrapper and unwrapper component.

Please note that the unwrapper serves as a Facade [GHVJ94] which subsequently calls all unwrapping functions that are adequate for this application (here: unpacking and signature verification). After the license has successfully passed all checks, it is considered to be *valid* (see Section 4.8) and the wrapper returns the license in plain text. The license is now forwarded to the rights expression interpreter. If the license fails one of the unwrapper checks, the license is invalid and the access requests is automatically denied.

The interpreter parses the license, extracts all relevant information, and builds a runtime model of the application–specific CoSa. The runtime model is shown in Figure 7.8. It consists of several objects, that are all aggregated by the instance co01 which is of the *Contract*[2]. The reminders of contract objects are py01 (type *Party*), re01 (type *Resource*), p01 (type *Permission*), and c01 (type *Permission*). The relations among the different runtime objects are expressed via the intrinsic attribute relations (see Section 4.6). For example, the party py01 is related to the objects contract co01 and permission p01 via the roles agg_parent and has_perm. Whereas the permission p01 is related to the objects co01, py01, re01, and c01 via the roles agg_parent, granted_to, refers_to, respectively has_constraint. For a complete list of possible roles, please refer to Section 6.3.2.

The runtime model may then be queried for the contract objects and their respective attributes, e.g. the unique ID of the beneficiary, his/her roles, the resources, the granted permissions to these resources, the constraints that apply to the granted permissions, etc. In the next step, the secure viewer component extracts the access control–relevant contract information from the runtime model of the contract to initialise the corresponding access control service. Finally, the secure viewer calls the method checkAccessRequest(..) to handle the access request of the beneficiary, i.e. it evaluates whether the requested access rights (display the resource

---

[2]This allows to delete all contract objects if the contract itself is erased

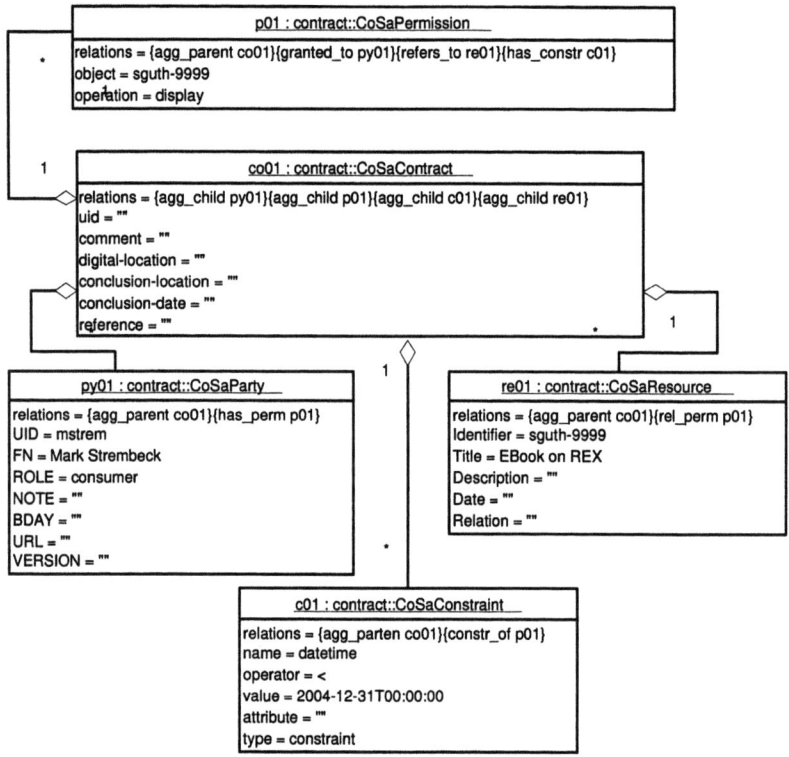

Figure 7.8: Runtime model of the DRM CoSa objects

sguth-9999) are granted in the ODRL license. The implementation, uses the xoRBAC access control service, which is described in Section 7.1. The access control mechanism grants the access requests, and therefore the secure viewer display the respective Ebook to M. Strembeck. The XOTcl code that implements the unwrapping, interpreting, and processing of the ODRL license is listed below.

```
#!/usr/local/bin/xotclsh

package require xoRBAC 0.6.1
package require rex::reInterpreter 0.1
package require rex::relContract 0.1
package require rex::reUnwrapper 0.1
```

205

```
Class SecureViewer

SecureViewer instproc requestAccess
    {subject operation object contract-location} {

  Unwrapper cc ${contract-location}
  set contract [cc unpack]
  set valid [cc verifySignature]

  if {($valid == "true")} {
    ODRLContract op $contract
    RightsManager rm
    set contracts [op getObjects CoSaContract]
    foreach c $contracts {
      set assets [op getAssets $c]
      set parties [op getRelatedObjects $c agg_child CoSaParty]
      set consumers [op selectObjects $parties ROLE "consumer"]

      foreach asset $assets {
      set assetId [op getAttributeValue $asset uid]

      foreach con $consumers {
        set conID [op getAttributeValue $con uid]
        rm createSubject $conID
        set rights [op getRelatedObjects $con rPerms]

        foreach r $rights {
          set right [op getAttributeValue $r name]
          rm createPermission "$right $assetId"
          rm subjectPermAssign "$conID" "$right $assetId"
          set constraints [op getRelatedObjects $r rConstr]

          foreach constr $constraints {
            set cname [op getAttributeValue $constr name]
            set cvalue [op getAttributeValue $constr value]
            set cop [op getAttributeValue $constr operator]
            if { $cname == "datetime"} {
            #this secure viewer can only handle time constraints

            rm createCondition $cname
            rm setConditionLeftOperand $cname "LocalhostSensor"
                                "lhsClock" "%Y-%m-%dT%H:%M:%S"
            rm setConditionOperator $cname $cop
            rm setConditionRightOperandAsConstant $cname $cvalue
            rm buildConditionScript $cname

            rm createContextConstraint "${cname}_$cvalue"
            rm addConditionToContextConstraint "$cname" "${cname}_$cvalue"
            rm addContextConstraintToPerm "${cname}_$cvalue" "$right $assetId"
              } else {
```

```
            op stdMsg "Other constraints than datetime_end are not provided"
    }}}}}}
    set result [rm grantAccess $subject $operation $object]
    if {$result == 1} {
     return 1
    } else {
     return 0
    }
  } else {
    cc ErrMsg "Contract ${contract-location} is not valid!"
    return 0
  }
}
SecureViewer sv set granted [sv requestAccess "mstrem" "play"
      "sguth-9999" ../odrl-instances/ODRL-EbookNo2.xml]
```

Evaluation of the Case Study

The rights expression interpreter is coded in approximately 3000 lines of XOTcl code, whereas the unwrapper component was implemented with approximately 500 lines of XOTcl code. Consequently, the be above 70 lines represent the functionality coded into the 3500 lines of the unwrapper and interpreter package (not including the xoRBAC code). The secure viewer above was run with the following performance:

party	user	system	real
elapsed time	0m0.310s	0m0.020s	0m0.359s

Here, the *user time* is the time the secure viewer is running, the *system time* is the time spend in system calls and real time is the total time the secure viewer has been running.

It is planned to make the source code of all components of the rights expression exchange framework freely available at the XOTcl web site[3] as well as on the web site of the ODRL initiative[4]. In terms of scalability, the framework can be extended to support a different application–specific CoSa and/or other metadata standards for resources and parties (e.g. LOM instead of Dublin Core) at low expense. The support of another rights expression language such as XrML, in return, would be more costly. In order to support the processing of more context constraints the mediator and/or the sensors of the access control service have to be extended.

---

[3]See: http://www.xotcl.org/
[4]See: http://odrl.net/

As mentioned earlier, it was a challenge to read the detailed semantics of the rights expression language ODRL from the written specification. Renato Iannella, the founder of the ODRL initiative, was very supportive in this matter. The development of rights expression language interpreters is at the very beginning. Apart from the work at hand, no design or comprehensive implementation of a rights expression language interpreter or rights expression generator is available. The most challenging issues of the generator implementation were: to design a comfortable, intuitive user interface, and to design a sensible policy for restricting the nestings of ODRL expressions.

Chapter 8
Conclusion and Future Work

As defined in Section 1.3, the goal of this doctoral thesis has been to develop methods and tools for exchanging and processing XML–based rights expressions, in particular electronic contracts. This section sums up the findings the thesis at hand delivers to the research community in this field. For details, please refer to the respective sections in the earlier chapters.

Methods for the exchange of rights expressions

The methods developed in the work at hand shall support the overall goal of creating rights expressions that are easily exchangeable with respect to their content. In other words, it is necessary to ensure that the receiver of a rights expression can understand its content and semantics as it has been intended of the sender. Within this thesis, the following methods have been developed to support the processability of electronic contracts in software services:

- *Tailoring electronic contracts.* The work provides an analysis of basic information objects that occur in rights expressions and their relations. Analysing electronic contracts and their structure is important for their reliable processing. Electronic contracts in particular can be processed in various usage scenarios, such as access control, accounting, customer relationship management, etc. For each usage scenario a specific agreement category has to be developed that describes objects that are relevant in this usage scenario. Accordingly, a process is introduced, that supports the tailored composition of agreement categories and thus electronic contracts depending on their later usage(s).

- *The generic Contract Schema (CoSa).* For a standardised processing of rights expressions and in particular of contracts, the concept of the generic *Contract Schema* (CoSa) has been developed. The CoSa serves as abstraction level for various rights expression languages or other representations of rights expression respectively contracts. With the concept of CoSa comes a generic API that allows to query all contract information represented in the CoSa format. The CoSa API stays consistent, also if the CoSa is application–specific.

- *The contract life cycle.* To build the bridge to the economic environment of exchanging and processing electronic contracts, the "contract life cycle" has been defined corresponding to the legal phases of contracts. With the help of the contract life cycle the developed methods and tools, as well as the defined terms have been classified to the respective life cycle phase. For example, the rights expression generator supports the *negotiation phase* of electronic contract, whereas the rights expression wrapper supports the offer placement as well as the offer confirmation, respectively the conclusion of the contract. Along the phases of the contract life cycle additionally the management issues are listed that are of importance, such as '*When is an electronic contract valid?*' or '*What technical means have to be applied, as soon as the electronic contract is executed (to avoid double-spending)?*'.

- *The enforceability matrix* that has been developed within this work helps to classify which information in rights expressions can be enforced as intended by the issuer. Three general criteria have been identified which facilitate the classification.

- *The rights expression communication model.* From Shannon's basic communication model a rights expression communication model has been derived that allows for particular needs of rights expression exchange. The developed model takes into account that rights expressions are exchanged in a platform–independent format and require security means to be transmitted safely. The four resulting stages that rights expression have to pass between sender and receiver are: rights expression generator, – wrapper, – unwrapper, and – interpreter.

Besides the above mentioned methods a system has been developed for the description and classification of DRM systems respectively their functionality. Additionally the perspectives of DRM systems have been identified that help understanding the various impacts (e.g. legal, social, technical, etc.) of DRM systems.

Tools for the exchange of rights expressions

The tools developed in this thesis implement the rights expression exchange model mentioned above. The tools have been developed as software components that can be used autonomously or in combination as *rights expression exchange framework*, comprising the four sub components (tools):

- The *rights expression generator* transforms original rights information of a DRM component or actor (e.g. a contract party) into code, resulting in a rights expression message. This message is formulated in a rights expression language. In return, the rights expression generator adopts the syntax and semantics of the open digital rights language (ODRL) version 1.1.

- The *rights expression wrapper and –unwrapper*. Depending on the usage scenario, various security services have to be applied to the rights expression before transmitting it, e.g. digitally signing and packing the rights expression in a secure container. These services are provided by the rights expression wrapper. The rights expression unwrapper is the complementary component to the rights expression wrapper. It unwraps and also unpacks the rights expression after transmission and provides for the extrinsic checking of the digital contract, such as checking the contract integrity, authenticating the rights expression sender, and verifying the digital signature of rights expression.

- The *rights expression interpreter* is an open and extensible tool for the interpretation of rights expressions for subsequent processing. The interpreter implements the concept of the generic CoSa. It is currently able to interpret ODRL instances and to transform them into an application–specific CoSa. The application–specific CoSa can then be queried for the rights information via the CoSa API. Thus, ODRL rights expressions are machine readable, and processable in various applications respectively usage scenarios.

Apart from this thesis, we are not aware of any other comprehensive study that supports methods and tools for the entire exchange task of rights expressions and for their subsequent processing. The developed methods are of generic nature and independent of any particular technology (e.g. a programming language or a rights expression language). The tools are prototype implementations that are open and extensible. They have well defined interfaces that assure its (re)use in various environments and ease its integration into existing systems. The implementations are coded in an appropriate programming language, reuse existing technology, and consider all relevant standards. The tools prove the correctness and usability of the introduced methods *tailored contract composition*, *rights expression communication model*, *CoSa*, and *enforceability*.

Taking everything into account, I come to the conclusion that the introduced methods and tools have the potential to bring forward current technology for the exchange of rights expressions (in particular the exchange of electronic contracts) in order to improve the interoperability of digital rights management systems and thus to quicken future electronic commerce.

Future Work

The work addresses a large number of subjects in the area of rights expression exchange and processing. Therefore, the future work to be done is equally broad. In the following paragraphes the fields of future work are mentioned that have my particular interest.

- The implementation of the rights expression interpreter has provided a detailed insight to rights expression languages. Due to the specialisation in certain applications respectively domains and to the few current usages of both XrML respectively MPEG 21 REL and ODRL they lack a well–defined data model for rights expressions and comprehensive, unambiguous, formal semantics. The lack of formal semantics considerably restricts the clarity of all existing RELs. Furthermore, today's RELs are not sufficiently designed for the later processing in software services. Consequently, the development of comprehensive formal semantics is a fundamental issue of future work in the field of

213

RELs. Also the introduced rights expression languages are not designed to support service level agreements. A subject to future work is an analysis, whether the predominant RELs are respectively should be able to express service level agreements.

- My investigations in the field of RELs have resulted in a close cooperation with Renato Iannella (founder of the ODRL initiative) and the people from ContentGuard who are developing ODRL respectively XrML. We intend to continue this fruitful cooperation in order to develop more sophisticated RELs. My participation in the ODRL initiative has resulted in the organisation of an international ODRL workshop in April 2004 in Vienna, where the leading researchers in the field of rights expression languages are going to meet to share their research findings. The achievements of the workshop will certainly have an impact on the next version of ODRL.

- Future work in this field will also be concerned with finding an adequate transport medium for electronic contracts, e.g. x509 certificates [IT93a]. Sandhu and Park have introduced *smart certificates* for attribute services on the web [PR99]. They use the extension field of X.509v3 certificates to bind attributes to a subject (party). In an other contribution Sandhu and Park [PS99] present an implementation where the extension field of a X.509v3 certificate is used to assign role information to a subject. Based on the subject's role information, web servers use roles instead of a user's identity for access control purposes. In my future work smart certificates, respectively the extension field of x509 certificates shall be investigated as a transport medium for electronic contracts with the help of a prototype implementation.

- This thesis addresses the perspectives of DRM systems and mentions their interrelation. For example, intentional perspective of a DRM system heavily influences its functions, respectively the technical implementation. Therefore, I consider it necessary to now investigate the concrete number of relations and their effect respectively the gravity of dependencies between the six perspectives. For example, a catalogue could be helpful that guides strategic and technic DRM system developers through system reengineering and change management.

- The thesis at hand addresses various security mechanisms that are required in DRM systems respectively in a rights expression exchange framework. Each of the security mechanism could be addressed and implemented in more conceptual detail. For a rights expression exchange framework, for example, the tamper resistance of the distinct components needs to be addressed for a concrete application. In this context it is also important to address the processing of electronic signatures in more detail. For example, for electronic contracts it has to be assured that the right people have signed the contract in the correct sequence.

- The most important challenge that future work should respond to is the creation of a standardised respectively globally unique vocabulary for rights and conditions, as they are available for individuals (e.g. x509) or resources (e.g. DOI), to improve the explicit semantics of rights expression languages. The MPEG 21 initiative is currently working on a uniform framework for the expression of rights in the course of developing part 6 of the MPEG 21 standard, a uniform data dictionary for rights expressions.

- To make sure that the application of electronic contracts is an absolute convenience for all future participant of e–commerce, is has to be taken care of the privacy matter in this context. It is important to design a process that ensures privacy for all contract parties, i.e. that addresses the management of privacy issues throughout the entire contract life cycle and covers the following issues: *Who decides on the contract content? Is the contract publicly available in whole or part?* etc.

- For the mapping from one rights expression language to the other, simple tables have been used in this thesis. For mapping the semantics from more than two rights expression languages a "rights ontology" needs to be developed.

Chapter 9

Appendix A:

ODRL Foundation Model and XML Schemata

This appendix comprises the data model and XML schemata of the Open Digital Rights Language (ODRL) [Ian02b]. The first XML schema describes the core language syntax of ODRL, the second XML schema is the standard extension of the language syntax, defining the ODRL vocabulary.

9.1 ODRL Foundation Model

Figure 9.1 shows the ODRL foundation model. The model consists of the three core entities assets, rights, and parties. With the reminder of entities, such as permission, agreement, context, condition, the three core entities can be described with more detail.

9.2 XML Schema of ODRL Syntax Version 1.1

```
<?xml version="1.0" encoding="UTF-8"?>
<xsd:schema
```

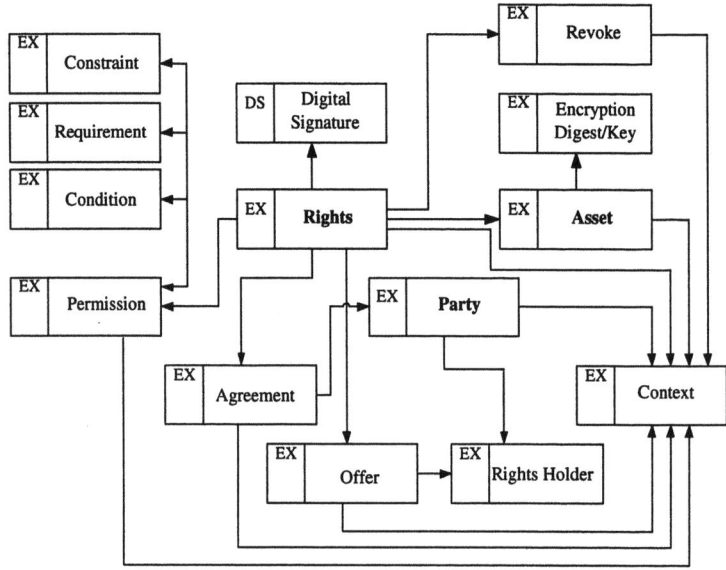

Figure 9.1: The foundation model of ODRL [Ian02b]

```
            targetNamespace="http://odrl.net/1.1/ODRL-EX"
            xmlns:o-ex="http://odrl.net/1.1/ODRL-EX"
            xmlns:xsd="http://www.w3.org/2001/XMLSchema"
            xmlns:ds="http://www.w3.org/2000/09/xmldsig#"
            xmlns:enc="http://www.w3.org/2001/04/xmlenc#"
            elementFormDefault="qualified" attributeFormDefault="qualified"
                                                               version="1.1">
<xsd:import namespace="http://www.w3.org/2000/09/xmldsig#"
   schemaLocation="http://www.w3.org/TR/2002/REC-xmldsig-core-20020212/
                                                    xmldsig-core-schema.xsd"/>
<xsd:import namespace="http://www.w3.org/2001/04/xmlenc#"
   schemaLocation="http://www.w3.org/Encryption/2001/Drafts/xmlenc-core/
                                                           xenc-schema.xsd"/>
     <xsd:element name="rights" type="o-ex:rightsType"/>
     <xsd:element name="offer" type="o-ex:offerAgreeType"/>
     <xsd:element name="agreement" type="o-ex:offerAgreeType"/>
     <xsd:complexType name="offerAgreeType">
         <xsd:choice minOccurs="0" maxOccurs="unbounded">
             <xsd:element ref="o-ex:context" minOccurs="0"
                                                    maxOccurs="unbounded"/>
             <xsd:element ref="o-ex:party" minOccurs="0"
                                                    maxOccurs="unbounded"/>
```

```
            <xsd:element ref="o-ex:asset" minOccurs="0"
                                          maxOccurs="unbounded"/>
            <xsd:element ref="o-ex:permission" minOccurs="0"
                                          maxOccurs="unbounded"/>
            <xsd:element ref="o-ex:constraint" minOccurs="0"
                                          maxOccurs="unbounded"/>
            <xsd:element ref="o-ex:requirement" minOccurs="0"
                                          maxOccurs="unbounded"/>
            <xsd:element ref="o-ex:condition" minOccurs="0"
                                          maxOccurs="unbounded"/>
        </xsd:choice>
</xsd:complexType>
<xsd:complexType name="rightsType">
    <xsd:complexContent>
      <xsd:extension base="o-ex:offerAgreeType">
        <xsd:choice minOccurs="0" maxOccurs="unbounded">
          <xsd:element ref="o-ex:revoke" minOccurs="0"
                                          maxOccurs="unbounded"/>
          <xsd:element ref="o-ex:offer" minOccurs="0"
                                          maxOccurs="unbounded"/>
          <xsd:element ref="o-ex:agreement" minOccurs="0"
                                          maxOccurs="unbounded"/>
          <xsd:element ref="ds:Signature" minOccurs="0"/>
        </xsd:choice>
        <xsd:attributeGroup ref="o-ex:IDGroup"/>
      </xsd:extension>
    </xsd:complexContent>
</xsd:complexType>
<xsd:element name="context" type="o-ex:contextType"/>
<xsd:element name="contextElement" abstract="true"/>
<xsd:complexType name="contextType">
    <xsd:choice minOccurs="0" maxOccurs="unbounded">
      <xsd:element ref="o-ex:context" minOccurs="0"
                                          maxOccurs="unbounded"/>
      <xsd:element ref="o-ex:contextElement" minOccurs="0"
                                          maxOccurs="unbounded"/>
    </xsd:choice>
    <xsd:attributeGroup ref="o-ex:IDGroup"/>
</xsd:complexType>
<xsd:complexType name="partyType">
    <xsd:choice minOccurs="0" maxOccurs="unbounded">
      <xsd:element ref="o-ex:context" minOccurs="0"/>
      <xsd:element ref="o-ex:rightsholder" minOccurs="0"
                                          maxOccurs="unbounded"/>
      <xsd:element ref="o-ex:party" minOccurs="0"
                                          maxOccurs="unbounded"/>
      <xsd:element ref="o-ex:container" minOccurs="0"
                                          maxOccurs="unbounded"/>
      <xsd:element ref="o-ex:asset" minOccurs="0"
                                          maxOccurs="unbounded"/>
```

```xml
        </xsd:choice>
        <xsd:attributeGroup ref="o-ex:IDGroup"/>
    </xsd:complexType>
    <xsd:element name="party" type="o-ex:partyType"/>
    <xsd:element name="rightsholder" type="o-ex:rightsHolderType"/>
    <xsd:element name="rightsHolderElement" abstract="true"/>
    <xsd:complexType name="rightsHolderType">
        <xsd:choice minOccurs="0" maxOccurs="unbounded">
            <xsd:element ref="o-ex:context" minOccurs="0"/>
            <xsd:element ref="o-ex:rightsHolderElement" minOccurs="0"
                                                       maxOccurs="unbounded"/>
            <xsd:element ref="o-ex:container" minOccurs="0"
                                              maxOccurs="unbounded"/>
        </xsd:choice>
        <xsd:attributeGroup ref="o-ex:IDGroup"/>
    </xsd:complexType>
    <xsd:complexType name="assetType">
        <xsd:choice minOccurs="0" maxOccurs="unbounded">
            <xsd:element ref="o-ex:context"/>
            <xsd:element ref="o-ex:inherit"/>
            <xsd:element name="digest">
                <xsd:complexType>
                    <xsd:choice minOccurs="0" maxOccurs="unbounded">
                        <xsd:element ref="ds:DigestMethod"/>
                        <xsd:element ref="ds:DigestValue"/>
                    </xsd:choice>
                </xsd:complexType>
            </xsd:element>
            <xsd:element ref="ds:KeyInfo"/>
        </xsd:choice>
        <xsd:attributeGroup ref="o-ex:IDGroup"/>
        <xsd:attribute name="type">
            <xsd:simpleType>
                <xsd:restriction base="xsd:NMTOKEN">
                    <xsd:enumeration value="work"/>
                    <xsd:enumeration value="expression"/>
                    <xsd:enumeration value="manifestation"/>
                    <xsd:enumeration value="item"/>
                </xsd:restriction>
            </xsd:simpleType>
        </xsd:attribute>
    </xsd:complexType>
    <xsd:element name="asset" type="o-ex:assetType"/>
    <xsd:complexType name="inheritType">
        <xsd:choice minOccurs="0" maxOccurs="unbounded">
            <xsd:element ref="o-ex:context" minOccurs="0"
                                            maxOccurs="unbounded"/>
        </xsd:choice>
        <xsd:attribute name="override" type="xsd:boolean" default="false"/>
        <xsd:attribute name="default" type="xsd:boolean" default="false"/>
```

```xml
</xsd:complexType>
<xsd:element name="inherit" type="o-ex:inheritType"/>
<xsd:element name="permission" type="o-ex:permissionType"/>
<xsd:element name="permissionElement" abstract="true"/>
<xsd:complexType name="permissionType">
    <xsd:choice minOccurs="0" maxOccurs="unbounded">
      <xsd:element ref="o-ex:context" minOccurs="0"
                                      maxOccurs="unbounded"/>
      <xsd:element ref="o-ex:permissionElement" minOccurs="0"
                                      maxOccurs="unbounded"/>
      <xsd:element ref="o-ex:container" minOccurs="0"
                                      maxOccurs="unbounded"/>
      <xsd:element ref="o-ex:constraint" minOccurs="0"
                                      maxOccurs="unbounded"/>
      <xsd:element ref="o-ex:sequence" minOccurs="0"
                                      maxOccurs="unbounded"/>
      <xsd:element ref="o-ex:requirement" minOccurs="0"
                                      maxOccurs="unbounded"/>
      <xsd:element ref="o-ex:condition" minOccurs="0"
                                      maxOccurs="unbounded"/>
      <xsd:element ref="o-ex:asset" minOccurs="0"
                                      maxOccurs="unbounded"/>
    </xsd:choice>
    <xsd:attribute name="exclusive" type="xsd:boolean" use="optional"/>
    <xsd:attributeGroup ref="o-ex:IDGroup"/>
</xsd:complexType>
<xsd:element name="constraint" type="o-ex:constraintType"/>
<xsd:element name="constraintElement" abstract="true"/>
<xsd:complexType name="constraintType">
    <xsd:choice minOccurs="0" maxOccurs="unbounded">
      <xsd:element ref="o-ex:constraint" minOccurs="0"
                                      maxOccurs="unbounded"/>
      <xsd:element ref="o-ex:constraintElement" minOccurs="0"
                                      maxOccurs="unbounded"/>
      <xsd:element ref="o-ex:container" minOccurs="0"
                                      maxOccurs="unbounded"/>
      <xsd:element ref="o-ex:sequence" minOccurs="0"
                                      maxOccurs="unbounded"/>
      <xsd:element ref="o-ex:context" minOccurs="0"
                                      maxOccurs="unbounded"/>
    </xsd:choice>
    <xsd:attributeGroup ref="o-ex:IDGroup"/>
    <xsd:attribute name="type" type="xsd:anyURI"/>
</xsd:complexType>
<xsd:element name="requirement" type="o-ex:requirementType"/>
<xsd:element name="requirementElement" abstract="true"/>
<xsd:complexType name="requirementType">
  <xsd:sequence minOccurs="0" maxOccurs="unbounded">
    <xsd:element ref="o-ex:context" minOccurs="0"
                                      maxOccurs="unbounded"/>
```

```xml
            <xsd:element ref="o-ex:requirementElement" minOccurs="0"
                                                       maxOccurs="unbounded"/>
            <xsd:element ref="o-ex:container" minOccurs="0"
                                                       maxOccurs="unbounded"/>
       </xsd:sequence>
       <xsd:attributeGroup ref="o-ex:IDGroup"/>
   </xsd:complexType>
   <xsd:element name="condition" type="o-ex:conditionType"/>
   <xsd:element name="conditionElement" abstract="true"/>
   <xsd:complexType name="conditionType">
       <xsd:sequence minOccurs="0" maxOccurs="unbounded">
           <xsd:element ref="o-ex:context" minOccurs="0"
                                                       maxOccurs="unbounded"/>
           <xsd:element ref="o-ex:conditionElement" minOccurs="0"
                                                       maxOccurs="unbounded"/>
           <xsd:element ref="o-ex:permission" minOccurs="0"
                                                       maxOccurs="unbounded"/>
           <xsd:element ref="o-ex:constraint" minOccurs="0"
                                                       maxOccurs="unbounded"/>
           <xsd:element ref="o-ex:container" minOccurs="0"
                                                       maxOccurs="unbounded"/>
           <xsd:element ref="o-ex:sequence" minOccurs="0"
                                                       maxOccurs="unbounded"/>
       </xsd:sequence>
       <xsd:attributeGroup ref="o-ex:IDGroup"/>
   </xsd:complexType>
   <xsd:complexType name="revokeType">
       <xsd:sequence minOccurs="0" maxOccurs="unbounded">
           <xsd:element ref="o-ex:context" minOccurs="0"
                                                       maxOccurs="unbounded"/>
       </xsd:sequence>
       <xsd:attributeGroup ref="o-ex:IDGroup"/>
   </xsd:complexType>
   <xsd:element name="revoke" type="o-ex:revokeType"/>
   <xsd:complexType name="sequenceType">
       <xsd:sequence>
           <xsd:element ref="o-ex:seq-item" maxOccurs="unbounded"/>
       </xsd:sequence>
       <xsd:attribute name="order" default="total">
           <xsd:simpleType>
               <xsd:restriction base="xsd:NMTOKEN">
                   <xsd:enumeration value="total"/>
                   <xsd:enumeration value="partial"/>
               </xsd:restriction>
           </xsd:simpleType>
       </xsd:attribute>
   </xsd:complexType>
   <xsd:element name="sequence" type="o-ex:sequenceType"/>
   <xsd:complexType name="containerType">
     <xsd:choice minOccurs="0" maxOccurs="unbounded">
```

```
        <xsd:element ref="o-ex:container" minOccurs="0"
                                          maxOccurs="unbounded"/>
        <xsd:element ref="o-ex:permission" minOccurs="0"
                                          maxOccurs="unbounded"/>
        <xsd:element ref="o-ex:permissionElement" minOccurs="0"
                                          maxOccurs="unbounded"/>
        <xsd:element ref="o-ex:constraintElement" minOccurs="0"
                                          maxOccurs="unbounded"/>
        <xsd:element ref="o-ex:conditionElement" minOccurs="0"
                                          maxOccurs="unbounded"/>
        <xsd:element ref="o-ex:requirementElement" minOccurs="0"
                                          maxOccurs="unbounded"/>
        <xsd:element ref="o-ex:rightsHolderElement" minOccurs="0"
                                          maxOccurs="unbounded"/>
        <xsd:element ref="o-ex:constraint" minOccurs="0"
                                          maxOccurs="unbounded"/>
        <xsd:element ref="o-ex:condition" minOccurs="0"
                                          maxOccurs="unbounded"/>
        <xsd:element ref="o-ex:sequence" minOccurs="0"
                                          maxOccurs="unbounded"/>
    </xsd:choice>
    <xsd:attribute name="type" default="and">
        <xsd:simpleType>
            <xsd:restriction base="xsd:NMTOKEN">
                <xsd:enumeration value="and"/>
                <xsd:enumeration value="in-or"/>
                <xsd:enumeration value="ex-or"/>
            </xsd:restriction>
        </xsd:simpleType>
    </xsd:attribute>
    <xsd:attributeGroup ref="o-ex:IDGroup"/>
</xsd:complexType>
<xsd:element name="container" type="o-ex:containerType"/>
<xsd:complexType name="seqItemType">
    <xsd:choice minOccurs="0" maxOccurs="unbounded">
        <xsd:element ref="o-ex:container" minOccurs="0"
                                          maxOccurs="unbounded"/>
        <xsd:element ref="o-ex:permission" minOccurs="0"
                                          maxOccurs="unbounded"/>
        <xsd:element ref="o-ex:permissionElement" minOccurs="0"
                                          maxOccurs="unbounded"/>
        <xsd:element ref="o-ex:constraintElement" minOccurs="0"
                                          maxOccurs="unbounded"/>
        <xsd:element ref="o-ex:conditionElement" minOccurs="0"
                                          maxOccurs="unbounded"/>
        <xsd:element ref="o-ex:requirementElement" minOccurs="0"
                                          maxOccurs="unbounded"/>
        <xsd:element ref="o-ex:rightsHolderElement" minOccurs="0"
                                          maxOccurs="unbounded"/>
        <xsd:element ref="o-ex:constraint" minOccurs="0"
```

```
                                          maxOccurs="unbounded"/>
       <xsd:element ref="o-ex:condition" minOccurs="0"
                                          maxOccurs="unbounded"/>
       <xsd:element ref="o-ex:sequence" minOccurs="0"
                                          maxOccurs="unbounded"/>
      </xsd:choice>
      <xsd:attribute name="number" type="xsd:integer" use="required"/>
    </xsd:complexType>
    <xsd:element name="seq-item" type="o-ex:seqItemType"/>
    <xsd:attributeGroup name="IDGroup">
       <xsd:attribute name="id" type="xsd:ID"/>
       <xsd:attribute name="idref" type="xsd:IDREF"/>
    </xsd:attributeGroup>
</xsd:schema>
```

9.3 XML Schema of ODRL Data Dictionary Version 1.1

```
<?xml version="1.0" encoding="UTF-8"?>
<xsd:schema targetNamespace="http://odrl.net/1.1/ODRL-DD"
       xmlns:o-ex="http://odrl.net/1.1/ODRL-EX"
       xmlns:o-dd="http://odrl.net/1.1/ODRL-DD"
       xmlns:xsd="http://www.w3.org/2001/XMLSchema"
       elementFormDefault="qualified" attributeFormDefault="qualified"
                                                          version="1.1">
    <xsd:import namespace="http://odrl.net/1.1/ODRL-EX"
       schemaLocation="http://odrl.net/1.1/ODRL-EX-11.xsd"/>
    <!-- Declare all the Permission Elements -->
    <xsd:element name="display" type="o-ex:permissionType"
                            substitutionGroup="o-ex:permissionElement"/>
    <xsd:element name="print" type="o-ex:permissionType"
                            substitutionGroup="o-ex:permissionElement"/>
    <xsd:element name="play" type="o-ex:permissionType"
                            substitutionGroup="o-ex:permissionElement"/>
    <xsd:element name="execute" type="o-ex:permissionType"
                            substitutionGroup="o-ex:permissionElement"/>
    <xsd:element name="sell" type="o-ex:permissionType"
                            substitutionGroup="o-ex:permissionElement"/>
    <xsd:element name="lend" type="o-ex:permissionType"
                            substitutionGroup="o-ex:permissionElement"/>
    <xsd:element name="give" type="o-ex:permissionType"
                            substitutionGroup="o-ex:permissionElement"/>
    <xsd:element name="lease" type="o-ex:permissionType"
                            substitutionGroup="o-ex:permissionElement"/>
    <xsd:element name="modify" type="o-ex:permissionType"
                            substitutionGroup="o-ex:permissionElement"/>
    <xsd:element name="excerpt" type="o-ex:permissionType"
```

```
                                    substitutionGroup="o-ex:permissionElement"/>
<xsd:element name="aggregate" type="o-ex:permissionType"
                                    substitutionGroup="o-ex:permissionElement"/>
<xsd:element name="annotate" type="o-ex:permissionType"
                                    substitutionGroup="o-ex:permissionElement"/>
<xsd:element name="move" type="o-ex:permissionType"
                                    substitutionGroup="o-ex:permissionElement"/>
<xsd:element name="duplicate" type="o-ex:permissionType"
                                    substitutionGroup="o-ex:permissionElement"/>
<xsd:element name="delete" type="o-ex:permissionType"
                                    substitutionGroup="o-ex:permissionElement"/>
<xsd:element name="verify" type="o-ex:permissionType"
                                    substitutionGroup="o-ex:permissionElement"/>
<xsd:element name="backup" type="o-ex:permissionType"
                                    substitutionGroup="o-ex:permissionElement"/>
<xsd:element name="restore" type="o-ex:permissionType"
                                    substitutionGroup="o-ex:permissionElement"/>
<xsd:element name="install" type="o-ex:permissionType"
                                    substitutionGroup="o-ex:permissionElement"/>
<xsd:element name="uninstall" type="o-ex:permissionType"
                                    substitutionGroup="o-ex:permissionElement"/>
<xsd:element name="save" type="o-ex:permissionType"
                                    substitutionGroup="o-ex:permissionElement"/>
<!-- Declare the Payment Element (used in Requirements) -->
<xsd:element name="payment">
    <xsd:complexType>
        <xsd:sequence>
            <xsd:element name="amount">
                <xsd:complexType>
                    <xsd:simpleContent>
                        <xsd:extension base="xsd:decimal">
                            <xsd:attribute name="currency"
                                    type="xsd:NMTOKEN" use="required"/>
                        </xsd:extension>
                    </xsd:simpleContent>
                </xsd:complexType>
            </xsd:element>
            <xsd:element name="taxpercent" minOccurs="0">
                <xsd:complexType>
                    <xsd:simpleContent>
                        <xsd:extension base="xsd:decimal">
                            <xsd:attribute name="code"
                                    type="xsd:NMTOKEN" use="required"/>
                        </xsd:extension>
                    </xsd:simpleContent>
                </xsd:complexType>
            </xsd:element>
        </xsd:sequence>
    </xsd:complexType>
</xsd:element>
```

```xml
<!-- Define the dataTypes used for Requirements using Payment element -->
<xsd:complexType name="feeType">
    <xsd:complexContent>
        <xsd:extension base="o-ex:requirementType">
            <xsd:sequence>
                <xsd:element ref="o-dd:payment"/>
            </xsd:sequence>
        </xsd:extension>
    </xsd:complexContent>
</xsd:complexType>
<!-- Declare all the Requirements Elements -->
<xsd:element name="prepay" type="o-dd:feeType"
                            substitutionGroup="o-ex:requirementElement"/>
<xsd:element name="postpay" type="o-dd:feeType"
                            substitutionGroup="o-ex:requirementElement"/>
<xsd:element name="peruse" type="o-dd:feeType"
                            substitutionGroup="o-ex:requirementElement"/>
<xsd:element name="accept" type="o-ex:requirementType"
                            substitutionGroup="o-ex:requirementElement"/>
<xsd:element name="register" type="o-ex:requirementType"
                            substitutionGroup="o-ex:requirementElement"/>
<xsd:element name="attribution" type="o-ex:requirementType"
                            substitutionGroup="o-ex:requirementElement"/>
<xsd:element name="tracked" type="o-ex:requirementType"
                            substitutionGroup="o-ex:requirementElement"/>
<!-- Declare all the RightsHolder Elements -->
<xsd:element name="fixedamount"
                            substitutionGroup="o-ex:rightsHolderElement">
    <xsd:complexType>
        <xsd:complexContent>
            <xsd:extension base="o-ex:rightsHolderType">
                <xsd:sequence>
                    <xsd:element ref="o-dd:payment"/>
                </xsd:sequence>
            </xsd:extension>
        </xsd:complexContent>
    </xsd:complexType>
</xsd:element>
<xsd:element name="percentage"
                            substitutionGroup="o-ex:rightsHolderElement">
    <xsd:simpleType>
        <xsd:restriction base="xsd:decimal">
            <xsd:minInclusive value="0"/>
            <xsd:maxInclusive value="100"/>
        </xsd:restriction>
    </xsd:simpleType>
</xsd:element>
<!-- Declare all the Context Elements -->
<xsd:simpleType name="uriAndOrString">
<xsd:union memberTypes="xsd:anyURI xsd:string"/>
```

```xml
</xsd:simpleType>
<xsd:element name="uid" type="o-dd:uriAndOrString"
                        substitutionGroup="o-ex:contextElement"/>
<xsd:element name="role" type="xsd:anyURI"
                        substitutionGroup="o-ex:contextElement"/>
<xsd:element name="name" type="xsd:string"
                        substitutionGroup="o-ex:contextElement"/>
<xsd:element name="remark" type="xsd:string"
                        substitutionGroup="o-ex:contextElement"/>
<xsd:element name="event" type="xsd:string"
                        substitutionGroup="o-ex:contextElement"/>
<xsd:element name="pLocation" type="xsd:string"
                        substitutionGroup="o-ex:contextElement"/>
<xsd:element name="dLocation" type="xsd:anyURI"
                        substitutionGroup="o-ex:contextElement"/>
<xsd:element name="reference" type="xsd:anyURI"
                        substitutionGroup="o-ex:contextElement"/>
<xsd:element name="version" type="xsd:string"
                        substitutionGroup="o-ex:contextElement"/>
<xsd:element name="transaction" type="xsd:string"
                        substitutionGroup="o-ex:contextElement"/>
<xsd:element name="service" type="xsd:anyURI"
                        substitutionGroup="o-ex:contextElement"/>
<xsd:element name="date" type="o-dd:dateType"
                        substitutionGroup="o-ex:contextElement"/>
<!-- Declare all the Constraint Elements -->
<xsd:element name="individual" type="o-ex:constraintType"
                        substitutionGroup="o-ex:constraintElement"/>
<xsd:element name="group" type="o-ex:constraintType"
                        substitutionGroup="o-ex:constraintElement"/>
<xsd:element name="cpu" type="o-ex:constraintType"
                        substitutionGroup="o-ex:constraintElement"/>
<xsd:element name="network" type="o-ex:constraintType"
                        substitutionGroup="o-ex:constraintElement"/>
<xsd:element name="screen" type="o-ex:constraintType"
                        substitutionGroup="o-ex:constraintElement"/>
<xsd:element name="storage" type="o-ex:constraintType"
                        substitutionGroup="o-ex:constraintElement"/>
<xsd:element name="memory" type="o-ex:constraintType"
                        substitutionGroup="o-ex:constraintElement"/>
<xsd:element name="printer" type="o-ex:constraintType"
                        substitutionGroup="o-ex:constraintElement"/>
<xsd:element name="software" type="o-ex:constraintType"
                        substitutionGroup="o-ex:constraintElement"/>
<xsd:element name="hardware" type="o-ex:constraintType"
                        substitutionGroup="o-ex:constraintElement"/>
<xsd:element name="spatial" type="o-ex:constraintType"
                        substitutionGroup="o-ex:constraintElement"/>
<xsd:element name="quality" type="o-ex:constraintType"
                        substitutionGroup="o-ex:constraintElement"/>
```

```xml
<xsd:element name="format" type="o-ex:constraintType"
                            substitutionGroup="o-ex:constraintElement"/>
<xsd:element name="unit" type="o-ex:constraintType"
                            substitutionGroup="o-ex:constraintElement"/>
<xsd:element name="watermark" type="o-ex:constraintType"
                            substitutionGroup="o-ex:constraintElement"/>
<xsd:element name="purpose" type="o-ex:constraintType"
                            substitutionGroup="o-ex:constraintElement"/>
<xsd:element name="industry" type="o-ex:constraintType"
                            substitutionGroup="o-ex:constraintElement"/>
<xsd:element name="count" type="xsd:positiveInteger"
                            substitutionGroup="o-ex:constraintElement"/>
<xsd:element name="range" substitutionGroup="o-ex:constraintElement">
    <xsd:complexType>
        <xsd:complexContent>
            <xsd:extension base="o-ex:constraintType">
                <xsd:sequence>
                    <xsd:element name="min" type="xsd:decimal"
                                                    minOccurs="0"/>
                    <xsd:element name="max" type="xsd:decimal"
                                                    minOccurs="0"/>
                </xsd:sequence>
            </xsd:extension>
        </xsd:complexContent>
    </xsd:complexType>
</xsd:element>
<xsd:element name="datetime" type="o-dd:dateType"
                            substitutionGroup="o-ex:constraintElement"/>
<xsd:simpleType name="dateAndOrTime">
    <xsd:union memberTypes="xsd:date xsd:dateTime"/>
</xsd:simpleType>
<xsd:complexType name="dateType">
    <xsd:complexContent>
        <xsd:extension base="o-ex:constraintType">
            <xsd:choice>
                <xsd:sequence>
                    <xsd:element name="start" type="o-dd:dateAndOrTime"
                                                    minOccurs="0"/>
                    <xsd:element name="end" type="o-dd:dateAndOrTime"
                                                    minOccurs="0"/>
                </xsd:sequence>
                <xsd:element name="fixed" type="o-dd:dateAndOrTime"
                                                    minOccurs="0"/>
            </xsd:choice>
        </xsd:extension>
    </xsd:complexContent>
</xsd:complexType>
<xsd:element name="accumulated" type="xsd:duration"
                            substitutionGroup="o-ex:constraintElement"/>
<xsd:element name="interval" type="xsd:duration"
```

```xml
                        substitutionGroup="o-ex:constraintElement"/>
<xsd:element name="recontext" type="xsd:boolean"
                        substitutionGroup="o-ex:constraintElement"/>
<!-- Transfer Permission is defined as a ContainerType to enable complete
expression of rights in the Constraint   -->
<xsd:element name="transferPerm" substitutionGroup="o-ex:container">
    <xsd:complexType>
        <xsd:complexContent>
            <xsd:extension base="o-ex:containerType">
                <xsd:attribute name="downstream" default="equal">
                    <xsd:simpleType>
                        <xsd:restriction base="xsd:NMTOKEN">
                            <xsd:enumeration value="equal"/>
                            <xsd:enumeration value="less"/>
                            <xsd:enumeration value="notgreater"/>
                        </xsd:restriction>
                    </xsd:simpleType>
                </xsd:attribute>
            </xsd:extension>
        </xsd:complexContent>
    </xsd:complexType>
</xsd:element>
</xsd:schema>
```

Chapter 10

Appendix B

10.1 CoSa Application Programming Interface

This chapter describes the application programming interface of the generic contract schema (CoSa). xoREL (see Section 6.3) is an implementation of the CoSa API. For each method, an individual description is provided, including its name, arguments, return value, and a short functional description. Remember that the abstract class RELContract actually provides the CoSa API and serves as Facade for xoREL. Therefore, the methods described below are the API methods of each implementation of the RELContract class, e.g. ODRLContract (see Section 6.3). Thus, the method prefix cosa represents an instance of any RELContract subclass (e.g. ODRLContract).

cosa getObjects object-type

- *Arguments:*

    - object-type: The name of any class that is a subclass of CoSaObject. The class has to be defined in the currently used contract schema.

- *Description:* This method returns all instances of the class *object-type* that are existing in the current runtime CoSa.

- *Return:* List of fully–qualified objects that are instances of *ClassName*, or -1 if current runtime CoSa does not comprise any instances of the class *ClassName*.

- *XOTcl-Example:* Defining a variable `allContracts` and initialising it with (a list of) instances of the class `CoSaContract` in the current runtime CoSa (value of `allContracts` e.g. ::contract001).

  `set allContracts [cosa getObjects CoSaContract]`

cosa getRelatedObjects cosaObject ?relation? ?object-type?

- *Arguments:*
    - `cosaObject`: The name of a fully–qualified CoSa object, from which related objects shall be retrieved.
    - `relation`: The way (i.e. role name) how the required objects stand in relation with the `cosaObject`.
    - `object-type`: The name of any class that is a subclass of `CoSaObject`. The class has to be defined in the currently used contract schema.

- *Description:*

  This method returns all instances related to `cosaObject`. The related instance can be optionally filtered by either the `relation`, or by the `object-type` of the related instance, or both.

- *Return:* List of fully–qualified objects where the specified conditions (`subject ?predicate? ?object-type?`) hold or -1 if no related objects are found, that meet the specified conditions.

- *XOTcl-Example:* Set the value of `allConstraints` to all CoSa objects that are related to a certain permission object (e.g. ::permission001) in the `constraint_by` manner:

  `set allConstraints [cosa ::permission001 constraint_by]`

  Get all objects related to the permission object:

  `set allObjects [cosa ::permission001]`

  Get all objects related to the permission object that are of the CoSa type `CoSaConstraint`.

  `set allRelations [cosa ::permission001 "" CoSaConstraint]`

  Get all objects related to the `CoSaContract` object ::contract001 in the `agg_child` manner that are of the CoSa type `CoSaResource`.

  `set allResources [cosa ::contract001 agg_child CoSaResource]`

cosa getRelations cosaObject ?relation?

- *Arguments:*

    - cosaObject: The name of a fully–qualified CoSa object, from which related objects shall be retrieved. The object has to be existent in the current runtime CoSa.

    - relation: The name of a relation (e.g. has_perm) between two CoSaObjects. All allowed relations can be found in the rdf–description of the contract schema.

- *Description:*

    This method returns the content of the relations attribute of of cosaObject. The content is a list of list of ⟨relation–type cosaObject⟩–pairs. If the argument relation is specified, all pairs are returned where relation equals relation-type.

- *Return:* List of ⟨relation–type, related–cosaObject⟩–pairs, optionally filtered by relation. If no relations are available or no fitting pair can be found -1 is returned.

- *XOTcl-Example:* Defining a variable allRelations and initialising it with all relations of the object ::contract001).

    set allRelations [cosa getRelations ::contract001]

cosa getRelationTypes cosaObject

- *Arguments:*

    - cosaObject: The name of a fully–qualified CoSa object. The object has to be existent in the current runtime CoSa.

- *Description:* This method returns a list of (all) relation–types of cosaObject.

- *Return:* List of relation-types (e.g. has_perm). If no relations are available -1 is returned.

- *XOTcl-Example:* Defining a variable allTypes and initialising it with all relation–types of the object ::contract001).

    set allTypes [cosa getRelationTypes ::contract001]

cosa getRelObjectTypes cosaObject ?relation?

- *Arguments:*
    - `cosaObject`: The name of a fully–qualified CoSa object. The object has to be existent in the current runtime CoSa.
    - `relation`: The name of a relation (e.g. has_perm). All allowed relations can be found in the RDF–description of the contract schema.
- *Description:* The method returns the object–types (classes) of all CoSa objects related to `cosaObject` optionally filtered for a given `relation`.
- *Return:* A list fully–qualified CoSa class names (e.g. ::CoSaResource) -1 is returned if no relation, respectively no adequate relation has been found in the current runtime CoSa.
- *XOTcl-Example:* Defining a variable `getROTypes` and initialising it with all object–types related to ::contract001 in the agg_child-manner).
  `set types [cosa getRelObjectTypes ::contract001 agg_child]`

cosa hasRelation cosaObject relation

- *Arguments:*
    - `cosaObject`: The name of a fully–qualified CoSa object. The object has to be existent in the current runtime CoSa.
    - `relation`: The name of a relation (e.g. has_perm). All allowed relations can be found in the RDF–description of the contract schema.
- *Description:* The method determines whether `cosaObject` has a relation to a specific other instance or not.
- *Return:* This method returns a boolean value (either 1 or 0). 1 is returned if `cosaObject` is related to other objects in the `relation`-manner. If no objects of this relations are available 0 is returned.
- *XOTcl-Example:* Defining a variable `has_relation` and initialising it with all relation–types of the object ::contract001).
  `set has_relation [cosa hasRelation ::contract001 agg_child]`

cosa getAllAttributes cosaObject

- *Arguments:*
  - `cosaObject`: The name of a fully-qualified CoSa object. The object has to be existent in the current runtime CoSa.

- *Description:* This method returns all attributes of `cosaObject`.

- *Return:* A list of attribute names (e.g. uid, name, etc.). Returns an empty list if the cosa object does not comprise any variables.

- *XOTcl-Example:* Defining a variable `allAtts` and initialising it with all variable names of the object `::resource001`).
  `set allAtts [cosa getAllAttributes ::contract001]`

cosa getAttributeValue cosaObject attribute

- *Arguments:*
  - `cosaObject`: The name of a fully-qualified CoSa object. The object has to be existent in the current runtime CoSa.
  - `attribute`: The name of an instance attribute (e.g. uid, name, relation). All valid instance variable names can be found in the RDF-description of the contract schema.

- *Description:* The method returns the value(s) of the `attribute` of `cosaObject`. For example,

- *Return:* Returns the value of the respective instance attribute or -1 if the attribute is not available with the respective `cosaObject`.

- *XOTcl-Example:* Defining an attribute `uniqueID` and initialising it with the value of the attribute `uid` of the cosa object `::party001`).
  `set uniqueID [cosa getAttributeValue ::party001 uid]`

cosa selectObjects list attribute ?value?

- *Arguments:*
  - `list`: A list of fully-qualified CoSa object names. The objects have to be existent in the current runtime CoSa.

- **attribute:** The name of an instance variable (e.g. uid, name, ROLE). All valid instance variable names can be found in the RDF–description of the contract schema.

- **value:** A possible value of **variable** (e.g. isbn-12344566, "S. Guth", "consumer").

- *Description:* With this method a list of cosa objects can be filtered with respect to a certain attribute, respectively its value.

- *Return:* Returns a list of cosa objects that hold **variable** (with optionally the respective **value**). An empty list is returned if no cosa object in **list** can be found that meets these conditions.

- *XOTcl–Example:* Defining a variable **consumers** and initialising it with all cosa objects that have the instance variable ROLE with the value **consumer**.

    set consumers [op selectObjects $parties ROLE consumer]

10.2 Extended CoSa Application Programming Interface

Although all contract information can be retrieved with the generic API, it might be desirable to extend the API by methods that cover some specific, frequent queries. The following methods have been added to the CoSa API. All methods are implemented with the core API described above, thus additional API methods can be coded easily.

cosa getContracts

- *Arguments:* –

- *Description:* Within a rights expression several contracts can be available. That is, for example, if an ODRL document comprises several <agreement> tags, respectively an XrML document comprises several <grant> tags. This method goes through the document and returns all contracts.

- *Return:* List of fully–qualifies CoSaContract objects, or -1 if actual CoSa does not comprise any contract objects.

235

cosa getAssets contracts

- *Arguments:* –

    – contracts: list of fully–qualified contract objects (e.g. ::odrlc001 ::odrlc002)

- *Description:* This method returns all assets that are subject to a contract in the list of contracts.

- *Return:* List of fully–qualifies CoSaResource objects, or -1 if actual CoSa does not comprise any assets that are aggregated to the respective contract object(s).

cosa getConsumers contract

- *Arguments:*

    – contract: a single, fully–qualified contract object (e.g. ::odrlc001)

- *Description:* Get list of all consumers that are related to the contract object

- *Return:* List of all fully–qualified CoSaParty object names, that are related to contract with the role "consumer".

cosa getRightsholders contract

- *Arguments:*

    – contract: a single, fully–qualified contract object (e.g. ::odrlc001)

- *Description:* Get list of all rightsholders that are related to the contract object

- *Return:* List of all fully–qualified CoSaParty object names, that are related to contract with the role "seller".

cosa getName anyObject

- *Arguments:*

    – anyObject: fully–qualified object name of any class that has been derived from the CoSaObject class.
    (e.g. ::odrlc001, ::odrlc001::permission001)

- *Description:* Get the value of the instance variable name of anyObject.

- *Return:* Name of instance variable name, e.g. *print* or *John Doe.*

cosa getUniqueID anyObject

- *Arguments:*

    - anyObject: fully–qualified object name of any class that has been derived from the CoSaObject class.
    (e.g. ::odrlc001, ::odrlc001::permission001)

- *Description:* Get the value of the instance variable uid of anyObject.

- *Return:* Name of instance variable uid, e.g. an URN, ISBN, or DOI.

cosa getConstraints permission

- *Arguments:*

    - permission: fully–qualified object name of a CoSaPermission object that has been derived from the CoSaObject class.
    (e.g. ::odrlc001::permission001)

- *Description:* Get all CoSaConstraint objects that are related to permission

- *Return:* List of CoSaConstraint objects.

cosa getDuties party

- *Arguments:*

    - party: fully–qualified object name of a CoSaParty object that has been derived from the CoSaObject class. (e.g. ::odrlc001::party001)

- *Description:* Get all CoSaDuty objects that are related to party

- *Return:* List of CoSaDuty objects.

cosa getPermissions party

- *Arguments:*
    - `party`: fully–qualified object name of a `CoSaParty` object that has been derived from the `CoSaObject` class. (e.g. ::odrlc001::party001)

- *Description:* Get all `CoSaPermission` objects that are related to party

- *Return:* List of `CoSaPermission` objects.

10.3 Wrapper / Unwrapper Application Programming Interface

This chapter describes the application programming interface of the rights expression wrapper respectively unwrapper component (see Section 6.4). For the methods, an individual description is provided, including its name, arguments, return value, and a short functional description. The class `Wrapper` and `Unwrapper` serves as Facade for the `reWrapper` respectively `reUnwrapper` component. Thus, the method prefix w and unw represents an instance of a `Wrapper` respectively `Unwrapper` class.

w init re

- *Arguments:*
    - `re`: the rights expression that shall be wrapped.

- *Description:* Creates a new wrapper instance and initiated it with a rights expression.

- *Return:* -

- *XOTcl-Example:*
    `Wrapper w $re-location`

w wrap ?key?

- *Arguments:*
    - `key`: the private key which shall be used for the digital signature if not the default key (e.g. platform key) is taken.

- *Description:* Invokes all methods, that are required to wrap the rights expression for a secure transport.

- *Return:* Returns **true**, if wrapping was successful.

- *XOTcl-Example:*

  `w $sign-key`

w hash re

- *Arguments:*

  - **re**: the rights expression of which a hash shall be created. The default hash algorithm is SHA1.

- *Description:* Create a hash of the document that was given as parameter **re**.

- *Return:* Returns a SHA1 hash.

- *XOTcl-Example:*

  `w hash $re`

w sign ?key?

- *Arguments:*

  - **key**: the private key which shall be used for the digital signature if not the default key (platform key) is taken.

- *Description:* Digitally signs the rights expression.

- *Return:* Returns the digital signature.

- *XOTcl-Example:*

  `w sign $re`

w pack re hash signature

- *Arguments:*
    - re: the rights expression that shall be exchanged.
    - hash: the hash of the rights expression given as parameter re.
    - signature: the digital signature or the rights expression given as parameter re.
- *Description:* Packs rights expression, hash, and signature in a zip archive.
- *Return:* Returns the archive including rights expression, hash, and signature.
- *XOTcl-Example:*
    w $re $hash $sig

unw init archive

- *Arguments:*
    - archive: the zip archive including rights expression, signature, and hash that shall be unwrapped.
- *Description:* Creates a new unwrapper instance and initiated it with the zip archive given as parameter archive.
- *Return:* Returns true if the instance was successfully created.
- *XOTcl-Example:*
    Unwrapper unw $archive

unw unwrap ?cert?

- *Arguments:*
    - cert: the certificate, including the public key, with which the digital signature shall be verified (if not the default certificate (e.g. platform certificate) is taken.
- *Description:* Invokes all methods, that are required to unwrap the rights expression for further processing.

- *Return:* Returns true, if the unwrapping was successful.
- *XOTcl-Example:*

  unw unwrap $cert

unw unpack

- *Arguments:* -
- *Description:* Unpacks the rights expression, the hash an the digital signature from the zip archive. The zip archive is retrieved from an instance variable of the current unwrapper object.
- *Return:* Returns the rights expression in plain text, and stores the signature and the hash as instance variables of the current unwrapper object.
- *XOTcl-Example:*

  unw unpack

unw verifySignature ?cert?

- *Arguments:*
  - cert: the certificate, including the public key, with which the digital signature shall be verified (if not the default certificate (e.g. platform certificate) is taken.
- *Description:* Verifies the digital signature of the rights expression.
- *Return:* Returns true if the signature is valid.
- *XOTcl-Example:*

  unw verifySignature $cert

unw hash ?re?

- *Arguments:*
  - re: the rights expression of which a hash shall be created. The default hash algorithm is SHA1.

- *Description:* Creates a hash of the rights expression that was given as parameter re. The rights expression can also be retrieved from an instance variable of the current unwrapper object.
- *Return:* Returns a SHA1 hash.
- *XOTcl-Example:*

  unw hash $re

unw verifyIntegrity re signedHash

- *Arguments:*
  - re: the rights expression of which the integrity shall be verified.
  - signedHash: the hash that has been delivered within the zip archive.
- *Description:* Creates a new hash from the rights expression given in parameter re and compares it with the hash that has been delivered within the zip archive.
- *Return:* Returns true if the two hashes are identical, i.e. the integrity has been approved.
- *XOTcl-Example:*

  unw verifySignature $re $hash

Bibliography

[AGT01]   A. Anagnostopoulos, M.T. Goodrich, and R. Tamassia. Persistent Authenticated Dictionaries and Their Applications. *Lecture Notes in Computer Science*, 2200:379 f, 2001.

[AK96]   R. Anderson and M. Kuhn. Tamper Resistance - a Cautionary Note. In *Proceedings of the 2nd USENIX Workshop on Electronic Commerce*, November 1996.

[Ame01]   American Bar Association (ABA). Click-Through Agreements: Strategies for Avoiding Disputes on Validity of Assent. White Paper, Business Law Section Working Group on Electronic Contracting Practices, http://www.abanet.org/, 2001.

[ASBA99] C. Avgerou, J. Siemer, and N. Bjorn-Andersen. The academic field of information systems in europe. *European Journal of Information Systems*, 8:2:136–153, 1999.

[Ass00]   Association of American Publishers, Inc. Digital Rights Management for Ebooks: Publisher Requirements. Technical Report, http://www.publishers.org/digital/drm.pdf, 2000.

[Bak98]   Y. Bakos. The Emerging Role of Electronic Marketplaces on the Internet. *Communications of the ACM*, 41 (8), August 1998.

[BBF+02] M. Bartel, J. Boyer, B. Fox, B. LaMacchia, and E. Simon. XML–Signature Syntax and Processing. W3 Consortium Recommendation, http://www.w3.org/TR/xmldsig–core/, February 2002.

[BG00] D. Brickley and R.V. Guha. Resource description framework (RDF) schema specification 1.0. http://www.w3.org/TR/rdf-schema/, March 2000. W3 Consortium Candidate Recommendation.

[Bir01] S. Bird. Machine-Readable Rights Information. White Paper, http://www.ldc.upenn.edu/sb/, October 2001.

[BJPW99] A. Beugnard, J.M. Jezequel, N. Plouzeau, and D. Watkins. Making Components Contract Aware. *IEEE Computer Magazine*, 32(7), July 1999.

[BKKM97] H.U. Buhl, W. Koenig, H. Krcmar, and P. Mertens. German Perspectives on Information Systems: Research Topics, Methodological Challenges, and Patterns of Exchange with IS Practice. In *International Conference on Information Systems (ICIS)*, pages 529–530, 1997.

[BL94] T. Berners-Lee. Universal resource identifiers in www. Internet Engineering Task Force (IETF), RFC 1630, http://www.w3.org/Adressing/rfc1630.txt, June 1994.

[BLFF96] T. Berners-Lee, R. Fielding, and H. Frystyk. Hypertext Transfer Protocol – HTTP/1.0. Internet Engineering Task Force (IETF) RFC 1945, Standards Track, http://www.ietf.org/rfcs/rfc1945.txt, May 1996.

[BLFM98] T. Berners-Lee, R. Fielding, and L. Masinter. Uniform Resource Identifiers (URI): Generic Syntax. Internet Engineering Task Force (IETF) RFC 2396, Standards Track, http://www.ietf.org/rfcs/rfc2396.txt, August 1998.

[BM01] P.V. Biron and A. Malhotra. eXtenisble Markup Language (XML) Schema Part 2: Datatypes. W3 Consortium Recommendation, http://www.w3.org/TR/xmlschema-2/, May 2001.

[Bor02] J. Bormans. MPEG–21 Requirements Version 1.3. Technical Report ISO/IEC JTC 1/SC 29/WG 11/ N5232, Moving Pictures Expert Group (MPEG), http://mpeg.telecomitalialab.com/, Oktober 2002.

[BPSMM00] T. Bray, J. Paoli, C.M. Sperberg–McQueen, and E. Maler. eXtenisble Markup Language (XML) 1.0 (Second Edition). W3 Consortium Candidate Recommendation, http://www.w3.org/TR/XML/Core/, October 2000.

[BR02] C. Barlas and G. Rust. MPEG 21 Part 6: Rights Data Dictionary. Technical Report ISO/IEC JTC 1/SC 29/WG 11/N4943, Moving Pictures Expert Group, http://mpeg.telecomitalialab.com/, July 2002.

[BRB99] A. Bertsch, K. Rannenberg, and H. Bunz. *Sicherheitsinfrastrukturen Grundlagen, Realisierungen, Rechtliche Aspekte, Anwendungen*, chapter Nachhaltige Überprüfbarkeit digitaler Signaturen, pages 39–50. Vieweg Verlag, Wiesbaden, 1999.

[CCL+03] C.N. Chong, R. Corin, Y.W. Law, S. Etalle, and P.H. Hartel. LicenseScript - A Novel Digital Rights Language. In *Proceedings of the International Workshop for Technology, Economy, Social and Legal Aspects of Virtual Goods*, May 2003.

[CD99] J. Clark and S. DeRose. XML Path Language (XPath). http://www.w3.org/TR/xpath, November 1999. W3 Consortium Recommendation.

[CM03] E. Christiaanse and M.L. Markus. Participation in Collaboration Electronic Marketplaces. In *Proceedings of the 36th Hawaii International Conference on System Sciences (HICSS), Hawaii, USA*, January 2003.

[Con00] ContentGuard Inc. Extensible rights Markup Language (XrML) Version 2.0. Technical Specification, http://www.xrml.org/, 2000.

[Cox94] B. Cox. Superdistribution? *Wired Magazine*, September 1994.

[CSW97] S.Y. Choi, C.O. Stahl, and A.B. Whinston. *The economics of electronic commerce*. Macmillan Technical Publishing, IN, USA, 1997.

[DA99] T. Dierks and C. Allen. The Transport Layer Security (TLS) Protocol, Version 1.0. Internet Engineering Task Force (IETF) RFC 2246, Standards Track, http://www.ietf.org/rfcs/rfc2246.txt, January 1999.

[Dav89] F.D. Davis. Perceived usefulness, perceived ease of use, and user acceptance of information technology. *MIS Quarterly*, pages 319 – 340, September 1989.

[Deu00a] Deutsches Bundesministerium der Justiz. *Bürgerliches Gesetzbuch (BGB) 1.*, chapter Fernabsatzgesetz (FernAbsG), pages 897–909. 2000.

[Deu00b] Deutsches Bundesministerium der Justiz. *Bürgerliches Gesetzbuch (BGB) 1.*, chapter Novelle des Signaturgesetzes (Entwurf eines Gesetzes über Rahmenbedingungen für elektronische Signaturen und zur Änderung weiterer Vorschriften), page 876. May 2000.

[Deu01a] Deutsches Bundesministerium der Justiz. Gesetz zum Elektronischen Geschäftsverkehr (EGG). *Bundesgesetzblatt*, 70:3721, December 2001.

[Deu01b] Deutsches Bundesministerium der Justiz. Gesetz zur Anpassung der Formvorschriften des Privatrechts und anderer Vorschriften an den modernen Rechtsgeschäftsverkehr. *Bundesgesetzblatt*, Juli 2001.

[Dig03] Digital World Services, LLC. The ADo^{2}RA System. White Paper, http://www.dwsco.com/ps_adora.html/, January 2003.

[DK01] J. Duhl and S. Kevorkian. Understanding DRM Systems. White Paper, IDC. Technical White Paper, http://www.intertrust.com/, 2001.

[Dub01] Dublin Core Metadata Initiative. Doublin Core Metadata Element Set, Version 1.1. http://dublincore.org/documents/dces/, 2001.

[DWW03] T. DeMartini, X. Wang, and B. Wragg. MPEG 21 – Part 5, Rights Expression Language. http://mpeg.tilab.com/workingdocuments/mpeg-21/rel/, March 2003.

[EJ01] D. Eastlake and P. Jones. Secure Hash Algorithm (SHA1). Internet Engineering Task Force (IETF) RFC 3174, Standards Track, http://www.ietf.org/rfcs/rfc3174.txt, September 2001.

[Eri01] J.S. Erickson. Information Objects and Rights Management. A Mediation–based Approach to DRM Interoperability. *D–Lib Magazine*, 7:4, 2001.

[Eri02]   J.S. Erickson. OpenDRM: A Standards Framework for Digital Rights Expression, Messaging and Enforcement. Technical Report, Hewlett Packard, http://xml.coverpages.org/ EricksonOpenDRM200020902.pdf, 2002.

[Eur97]   European Union. Directive 97/7/EC of the European Parliament and of the Council on the protection of consumers in respect of distance contracts. *Official Journal of the European Communities*, May 1997.

[Eur99]   European Union. Directive 1999/93/EC of the European Parliament and of the Council on a Community Framework for Electronic Signatures. *Official Journal of the European Communities*, December 1999.

[Eur00]   European Union. Directive 2000/31/EC of the European Parliament and of the Council on certain legal aspects of information society services, in particular electronic commerce, in the Internal Market (Directive on electronic commerce). *Official Journal of the European Communities*, June 2000.

[Fed02]   H. Federrath. Scientific evaluation of drm systems. In *Digital Rights Management (DRM) Conference, Berlin/Germany*, January 2002.

[FFSS01]   J. Feigenbaum, M.J. Freedman, T. Sander, and A. Shostack. Security and privacy in digital rights management. In *Proceedings of ACM CCS-8 Workshop DRM: Privacy Engineering for Digital Rights Management Systems*, pages 76–106, 2001.

[FGM+99]   R. Fielding, J. Gettys, J. Mogul, H. Frystyk, L. Masinter, P. Leach, and T. Berners-Lee. Hypertext Transfer Protocol – HTTP/1.1. Internet Engineering Task Force (IETF) RFC 2616, Standards Track, http://www.ietf.org/rfcs/rfc2616.txt, June 1999.

[FHBH+99]   J. Franks, P. Hallam-Baker, J. Hostetler, S. Lawrence, P. Leach, A. Luotonen, and L. Stewart. HTTP Authentication: Basic and Digest Access Authentication. Internet Engineering Task Force (IETF) RFC 2617, Standards Track, http://www.ietf.org/rfcs/rfc2617.txt, June 1999.

[FKK96]   A.O. Freier, P. Karlton, and P.C. Kocher. The SSL Protocol, Version 3.0. Internet Draft, March 1996.

[FKT+99]   K. Fujimura, H. Kuno, M. Terada, K. Matsuyama, Y. Mizuno, and J. Sekine. Digital–ticket–controlled digital ticket circulation. In *Proceedings of the 8th USENIX Security Symposium, Washington D.C./USA*, page 229238, August 1999.

[FNS99]   K. Fujimura, Y. Nakajima, and J. Sekine. Xml–ticket: Generalized digital ticket definition language. NTT Information Sharing Platform Laboratories, http://www.w3.org/DSig/signed–SML99/pp% slashNTT_xml_ticket.html, 1999.

[Fra99]   U. Frank. *Wirtschaftsinformatik und Wissenschaftstheorie: Bestandsaufnahme und Perspektiven.*, chapter Zur Verwendung formaler Sprachen in der Wirtschaftsinformatik: Notwendiges Merkmal eines wissenschaftlichen Angspruchs oder Ausdruck eines übertriebenen Szientismus?, pages 127–160. Wiesbaden: Gabler, 1999.

[FS92]   J. Farrell and G. Saloner. Converters, Compatibility, and the Control of Interfaces. *Journal of Industrial Economics*, 40:1:9–35, 1992.

[FSG+01]   D.F. Ferraiolo, R. Sandhu, S. Gavrila, D.R. Kuhn, and R. Chandramouli. Proposed NIST Standard for Role–Based Access Control. *ACM Transactions on Information and System Security*, 4(3), August 2001.

[GHM00]   A. Goodchild, C. Herring, and Z. Milosevic. Businesss Contracts for B2B. In *Proceedings of the 9th International Conference Information Systems (ISD), Kristiansand, Norway*, August 2000.

[GHVJ94]   E. Gamma, R. Helm, J. Vlissides, and R. Johnson. *Design Patterns: Elements of Reusable Object Oriented Software.* Addison Wesley Longman, Inc., October 1994.

[Gil93]   G. Gilder. Metcalfe's law and legacy. *Forbes ASAP*, September 1993.

[GK02]   S. Guth and E. Koeppen. Electronic Rights Enforcement in Leaning Media. In *Proceedings of the IEEE International Conference on Advanced Learning Technologies, Kazan/Russland*, September 2002.

249

[GNZ00] M. Goedicke, G. Neumann, and U. Zdun. Design and Implementation Constructs for the Development of Flexible, Component-Oriented Software Architectures. In *Proceeding of the Second International Symposium on Generative and Component-Based Software Engineering, Erfurt, Germany*, 2000.

[GP03] E.E. Grandon and J.M. Pearson. Perceived Strategic Value and Adoption of Electronic Commerce: Am Empirican Study of Small and Medium Sized Businesses. In *Proceedings of the 36th Hawaii International Conference on System Sciences (HICSS)*, January 2003.

[GSSG00] M. Gisler, K. Stanoevska-Slabeva, and M. Greunz. Legal Aspects of Electronic Contracts. In *Proceeding of the Conference for Infrastructures for Dynamic Business-to-Business Service Outsourcing (IDSO'00), Stockholm, Sweden*, June 2000.

[GSSS00] M. Greunz, B. Schopp, , and K. Stanoevska-Slabeva. Supporting Market Transaction through XML Contracting Containers. In *Proceedings of the Americas Conference on Information Systems (AMCIS)*, August 2000.

[GSZ03] S. Guth, B. Simon, and U. Zdun. A Contract and Rights Management Framework Design for Interacting Brokers. In *Proceedings of the 36th Hawaii International Conference on System Sciences (HICSS)*, January 2003.

[Gut03] S. Guth. *Digital Rights Management: Technological, Economical, Legal, and Political Aspects*, chapter A Sample DRM System, pages 150–161. Springer, November 2003.

[GWW01] C. Gunter, S. Weeks, and A. Wright. Models and Languages for Digital Rights, Technical Report. White Paper, Intertrust Star Lab, 2001.

[HE96] L.M. Hitt and E.Brynjolfsson. Productivity, business profitability, and consumer surplus: Three different measures of information technology value. *MIS Quarterly*, pages 121 – 142, June 1996.

[HF98] T. Howes and F.Dawson. vCard MIME Directory Profile. http://www.ietf.org/rfc/rfc2426.txt, September 1998. Internet Engineering Task Force (IETF), RFC 2426.

[Hol99] A. Holl. *Wirtschaftsinformatik und Wissenschaftstheorie: Bestandsaufnahme und Perspektiven.*, chapter Empirische Wirtschaftsinformatik und evolutionäre Erkenntnistheorie, pages 163 – 207. Wiesbaden: Gabler, 1999.

[Ian01] R. Iannella. Digital Rights Management (DRM) Architectures. *D-Lib Magazine*, 7/7, June 2001.

[Ian02a] R. Iannella. Colis ODRL Metadata Profile. Technical Report, IPR Systems, http://www.iprsystems.com/, August 2002.

[Ian02b] R. Iannella. Open Digital Rights Language (ODRL), Version 1.1. Technical Specification, ODRL Initiative, http://odrl.net, August 2002.

[Ian03a] R. Iannella. ANTA LOX Technical Functional Requirements. Technical Report, IPR Systems, http://www.iprsystems.com/, September 2003.

[Ian03b] R. Iannella. Mobile Digital Rights Management. Technical Report, IPR Systems, http://www.iprsystems.com/, October 2003.

[Ian03c] R. Iannella. Trading Learning Objects. In *Proceedings of the EDUCAUSE in Australasia Conference, Adelaide/Australia*, May 2003.

[IBD95] A.L. Iacovou, I. Benbasat, and A. Dexter. Electronic data interchange and small organizations: Adoption and impact of technology. *MIS Quarterly*, pages 465–485, December 1995.

[IBM02] IBM Corporation. Emms software suite. White Paper, http://www.ibm.com/software/data/emmsash, November 2002.

[IEE02] IEEE Learning Technology Standards Committee (LTSC). Draft Standard for Learning Object Metadata (LOM), Final Draft, IEEE 1484.12.1–2002. http://ltsc.ieee.org/doc/wg12/LOM_1484_12_1_v1_Final_Draft.pdf, July 2002.

[ISO86] ISO - International Organisation for Standardization. International Standard ISO 8879, Information Processing - Text and office systems - Standard Generalized Markup Language (SGML). available at: http://www.iso.ch/, June 1986.

[ISO92] ISO - International Organisation for Standardization. International Standard IS 2108:1992, International Standard Book Numbering (ISBN). available at: http://www.iso.ch/, 1992.

[ISO95] ISO - International Organisation for Standardization. International Standard IS 11179-5, Information Technology Specification and Standardization of Data Elements, Part 5: Naming and Identification Principles for Data Elements. available at: http://www.iso.ch/, 1995.

[ISO98] ISO - International Organisation for Standardization. International Standard IS 3297:1998, International Standard Serial Number (ISSN). available at: http://www.iso.ch/, 1998.

[IT93a] ITU-T. ITU-T Recommendation X.500: Information Technology–Open Systems Interconnection–The Directory: Overview of Concepts, Models and Services, 1993.

[IT93b] ITU-T. ITU-T Recommendation X.509: Information Technology – Open Systems Interconnection – The Directory: Authentication Framework, 1993.

[Kap96] M.A. Kaplan. IBM Cryptolopes, SuperDistribution and Digital Rights Management. Technical Report, IBM Corporation, 1996.

[KF02] G. Kerscher and J. Fruchterman. The Soundproof Book: Exploration of Rights conflict and Access to Commercial EBooks for People with Disabilities. White Paper, DAISY Consortium and Benetech Initiative, available at: http://www.abanet.org/, August 2002.

[KGV99] M. Koetsier, P. Grefen, and J. Vonk. *Contract Model*. Technical Report, CrossFlow – EC Research Project, http://www.crossflow.org/public/pubdel/D4b.pdf, August 1999.

[KGV00] M. Koetsier, P. Grefen, and J. Vonk. Contracts for Cross-Organizational Workflow Management. In *Proceedings of the 1st International Conference on Electronic Commerce and Web Technologies, London, UK*, pages 110–121, 2000.

[KHvP95] W. Koenig, A. Heinzl, and A. von Poblotzki. Die zentralen forschungsgegenstände der wirtschaftsinformatik. *Wirtschaftsinformatik*, 37, 1995.

[KKK+95] H. Krcmar, W. Koenig, K. Kurbel, D. B. Pressmar, A. Scheer, and W. Stucky. Panel: Current Research and Practice in Information Systems in Germany. In *Third European Conference on Information Systems, Athens, Greece*, pages 1295–1297, 1995.

[KKL+02] A. Keller, G. Kar, H. Ludwig, A. Dan, and J.L. Hellerstein. Managing Dynamic Services: A Contract Based Approach to a Conceptual Architecture. In *Proceedings of the 8th IEEE/IFIP Network Operations and Management Symposium (NOMS)*, April 2002.

[KL89] S. Kent and J. Linn. Privacy enhancement for internet electronic mail: Part II – certificate–based key management; RFC 1114. *Internet Request for Comments*, (1114), 1989.

[KL02] A. Keller and H. Ludwig. Defining and Monitoring Service Level Agreements for dynamic e–Business. In *Proceedings of the 16th System Administration Conference (LISA), Philadelphia, USA*, November 2002.

[KM00] D. Konstantas and J.H. Morin. Agent–based Commercial Dissemination of Electronic Information. *Computer Networks. The International Journal of Computer and Telecommunications Networking*, pages 753–765, May 2000.

[Kop99] O. Koppius. Dimensions of Intangible Goods. In *Proceedings of the 32nd Hawaii International Conference on System Sciences (HICSS)*, January 1999.

[KS94] M.L. Katz and C. Shapiro. System competition and network effects. *Journal of Economic Perspectives*, 8:2:93–115, 1994.

[LA03] J. Loewer and R. Ade. tDOM (DOM Implementation). available at: http://www.tdom.org/, 2003.

[Lan81] C.E. Landwehr. Formal Models for Computer Security. *ACM Computing Surveys*, 13(3), September 1981.

[LDF+02] H. Ludwig, A. Dan, R. Franck, A. Keller, and R.P. King. Web Service Level Agreement (WSLA) Language Specification. Technical Report, IBM Corporation, July 2002.

[Les01] L. Lessig. *The Future of Ideas*. Random House, Inc., 2001.

[LHM95]   F. Lehner, K. Hildebrand, and R. Maier. *Wirtschaftsinformatik: Theoretische Grundlagen.* Hanser, 1995.

[LMSZ00]   A.L. Lederer, D.J. Maupin, M.P Sena, and Y. Zhuang. The technology addeptance model and the World Wide Web. *Decision Support Systems*, 29:269–282, 2000.

[LS99]   O. Lassila and R.R. Swick. Resource description framework (RDF) model and syntax specification. http://www.w3.org/TR/REC-rdf-syntax/, February 1999. W3 Consortium Recommendation.

[LTM+00]   D. Lie, C. Thekkath, M. Mitchell, P. Lincoln, D. Boneh, J. Mitchell, and M. Horowitz. Architectural Support for Copy and Tamper Resistant Software. In *Proceedings of the 6th International Conference on Architectural Support for Programming Languages and Operating Systems*, November 2000.

[MAR03]   MARC Advisory Committee. MARC Code List for Relators, Sources, Description Conventions. http://www.loc.gov/marc/relators/, January 2003.

[MB95]   Z. Milosevic and A. Bond. Electronic Commerce on the Internet: What is Still Missing? In *Proceedings of the 5th Conf. of the Internet Society, Honolulu, USA*, pages 245–254, June 1995.

[MB03]   L.G. Meredith and S. Bjord. Contracts and Types. *Communications of the ACM*, 46:10:41–47, October 2003.

[MH02]   P. Mertens and L.J. Heinrich. *Entwicklungen der Betriebswirtschaftslehre*, chapter Wirtschaftsinformatik – Ein interdisziplinäres Fach setzt sich durch, pages 476–489. Schäffer-Poeschel Verlag Stuttgart, 2002.

[Mic03]   Microsoft Corporation. Understanding how Windows Media Rights Manager Works. White Paper, http://msdn.microsoft.com/library/, 2003.

[MJP02]   Z. Milosevic, A. Jsang, and M.-A. Patton. Discretionary Enforcement of Electronic Contracts. In *Proceedings of the 6th IEEE International Enterprise Distributed Object Computing Conference (EDOC) Lausanne, Switzerland*, August 2002.

[MN93] G. Medvinsky and C. Neuman. NetCash: A Design for Practical Electronic Currency on the Internet. In *Proceedings of the ACM Conference on Computer and Communications Security (CCS)*, November 1993.

[Moa97] S. Moats. Urn syntax RFC 2141. *IETF, Network Working Group: Standards Track*, May 1997.

[MSM01] M. Morciniec, M. Salle, and B. Manahan. Towards Regulating Electronic Communities with Contracts. White Paper, Hewlett Packard Laboratories Bristol, May 2001.

[Nat00] National Information Standards Organization (NISO). Syntax for the Digital Object Identifier. http://www.niso.org/standards/, December 2000.

[NCL+03] S. Neal., J. Cole, P. F. Linington, Z. Milosevic, S. Gibson, and S. Kulkarni. Identifying requirements for Business Contract Language: a Monitoring Perspectiv. In *Proceedings of the 7th IEEE International Enterprise Distributed Object Computing Conference (EDOC), Brisbane, Australia*, September 2003.

[Ney01] E. Neylon. First Steps in an Information Commerce Economy. *D–Lib Magazine*, 7, January 2001.

[NJRW01] H.A. Napier, P.J. Judd, O.N. Rivers, and S.W. Wagner. *Creating a Winning E-business*. Course Technology, Cambridge, MA, USA, 2001.

[Nok01] Nokia Mobile Phones. Digital Rights Management and Superdistribution of Mobile Content. White Paper, http://www.nokia.com/, 2001.

[NS01] G. Neumann and M. Strembeck. Design and Implementation of a Flexible RBAC–Service in an Object–Oriented Scripting Language. In *Proceedings of the 8th ACM Conference on Computer and Communications Security (CCS)*, November 2001.

[NS03a] G. Neumann and M. Strembeck. An Approach to Engineer and Enforce Context Constraints in an RBAC Environment. In *Proceedings of the 8th ACM Symposium on Access Control Models and Technologies (SACMAT)*, June 2003.

[NS03b] G. Neumann and M. Strembeck. An approach to engineer and enforce context constraints in an rbac environment. In *Proceedings of 8th ACM Symposium on Access Control Models and Technologies (SACMAT), Como, Italy*, June 2003.

[NZ99a] G. Neumann and U. Zdun. Enhancing object-based system composition through per-object mixins. In *Proceedings of Asia–Pacific Software Engineering Conference (APSEC), Takamatsu/Japan*, December 1999.

[NZ99b] G. Neumann and U. Zdun. Implementing object–specific design patterns using per–object mixins. In *Proceedings of Second Nordic Workshop on Software Architecture (NOSA), Ronneby/Sweden*, August 1999.

[NZ00a] G. Neumann and U. Zdun. High-Level Design and Architecture of an HTTP-Based Infrastructure for Web Applications. *World Wide Web Journal*, 3(1), 2000.

[NZ00b] G. Neumann and U. Zdun. XOTcl, an Object–Oriented Scripting Language. In *Proceedings of Tcl2k: 7th USENIX Tcl/Tk Conference*, February 2000.

[oAPI01] Association of American Publishers Inc. Digital Rights Management for Ebooks. http://www.publishers.org/, January 2001.

[Oct02] Octalis S.A. Custom Digital Rights Language (CDRL), Version 2.6. Technical Specification, http://octalis.com/R+D/rd.htm, June 2002.

[OLP02] A. Osterwalder, S.B. Lagha, and Y. Pigneur. An ontology for developing e–business models. In *Proceedings of the International Conference on Decision Making and Decision Support in the Internet Age (DSIage), Ireland*, July 2002.

[Ous94] J.K. Ousterhout. *Tcl and the Tk Toolkit*. Addison–Wesley, 1994.

[Pet03] F.A.P. Petitcolas. *Digital Rights Management: Technological, Economical, Legal, and Political Aspects*, chapter Digital Watermarking, pages 81–92. Springer, November 2003.

[PJ02]   A. Powell and P. Johnston. Guidelines for implementing Dublin Core in XML. Technical Report, Dublin Core Metadata Initiative, http://dublincore.org/, December 2002.

[Pos80]   J. Postel. User Datagram Protocol (UDP). Internet Engineering Task Force (IETF), Internet Requests for Comments, No. 768, August 1980.

[Pos81a] J. Postel. Internet Protocol (IP). Internet Engineering Task Force (IETF), Internet Requests for Comments, No. 791, September 1981.

[Pos81b] J. Postel. Transmission Control Protocol (TCP). Internet Engineering Task Force (IETF), Internet Requests for Comments, No. 793, September 1981.

[PR99]   J.S. Park and R.Sandhu. Smart Certificates: Extending X.509 for Secure Attribute Services on the Web. In *Proceedings of 22nd National Information Systems Security Conference (NISSC)*, October 1999.

[PS99]   J.S. Park and R. Sandhu. RBAC on the Web by Smart Certificates. In *Proceedings of the ACM Workshop on Role-Based Access Control*, October 1999.

[PS02a]   J. Park and R. Sandhu. Originator Control in Usage Control. In *Proceedings of the 3rd International Workshop on Policies for Distributed Systems and Networks*, June 2002.

[PS02b]   J. Park and R. Sandhu. Towards Usage Control Models: Beyond Traditional Access Control. In *Proceedings of the 7th ACM Symposium on Access Control Models and Technologies (SACMAT)*, June 2002.

[RB99]   G. Rust and M. Bride. The <indecs> Metadata Model. Technical Specification, MUZE - EDItEUR, http://www.editeur.org/, July 1999.

[Rig02]   Rightscom Ldt. A Standard Rights Data Dictionary. White Paper, The <indecs>rdd Consortium, http://www.rightscom.com/, May 2002.

[Rig03]   Rightscom Ldt. The MPEG–21 Rights Expression Language. Technical Report, http://www.rightscom.com/, July 2003.

[Riv97]   L.R. Rivest. *Financial Cryptography*, chapter Electronic Lottery Tickets as Micropayments, page 307314. Springer Verlag, November 1997.

[RTM02]   B. Rosenblatt, B. Trippe, and S. Mooney. *Digital Rights Management: Business and Technology*. M&T Books, New York/ USA, 2002.

[RvdV04]   S. Royer and R. van der Velden. Economics, E-Commerce and Strategy Development - Resources and Rent Creation for Digital Good Providers in the Internet. *International Journal of Management and Decision Making*, 2004.

[Sac00]   M. Sachs. Electronic Trading–Partner Agreement for E–Commerce. Pre–submission Draft, Version: 1.0.3. IBM T. J. Watson Research Center, USA, http://www.ibm.com/developer/xml/tpaml/tpaspec.pdf, January 2000.

[San96]   R.S. Sandhu. Roles versus Groups. In *Proceedings of ACM Workshop on Role–Based Access Control, Part I, MD,USA*, 1996.

[SB02]   B. Shand and J. Bacon. Policies in Accountable Contracts. In *Proceedings of the 3rd International Workshop on Policies for Distributed Systems and Networks*, June 2002.

[SCFY96]   R.S. Sandhu, E.J. Coyne, H.L. Feinstein, and C.E. Youman. Role–based access control models. *IEEE Computer*, 29(2), February 1996.

[Sch71]   W. Schramm. *The Nature of Communication Between Humans. In: The Process and Effects of Mass Communication*. University of Illinois Press, 1971.

[Sch79]   B. Schneier. *Applied Cryptography*, volume Second Edition. John Wiley & Sons, 1979.

[Sch03]   C. Schinagl. Erfinderleitfaden: Hinweise für Entwickler und Erfinder in der angewandten Forschung und Enticklung zum Thema Patentrecht und Patentrecherchen. Technical Report, Johanneum Research, http://www.joanneum.at/, September 2003.

[SD00] P. Samualson and R. Davis. THE DIGITAL DILEMMA: A Perspective on Intellectual Property in the Information Age. In *Proceedings of the 28th Annual Telecommunications Policy Research Conference, Virginia, USA*, September 2000.

[SDN+00] M. Sachs, A. Dan, T. Nguyen, R. Kearney, H. Shaikh, and D. Diaz. Executable Trading–Partner Agreements in Electronic Commerce. Technical Report, IBM Cooperation, http://www.ibm.com/developer/xml/tpaml/tpapaper.pdf, 2000.

[Sha48] C. E. Shannon. A mathematical theory of communication. *Bell System Technical Journal*, 27:379–423, and 623–656, July and October 1948.

[SNS88] J.G. Steiner, C. Neuman, and J.I. Schiller. Kerberos: An Authentication Service for Open Network Systems. In *Proc. of the USENIX Winter Conference*, February 1988.

[SP00] G.P. Schneider and J.T. Perry. *Electronic Commerce*. Course Technology, Cambridge, MA, USA, 2000.

[SS02] G. Saloner and A.M. Spence. *Creating and Capturing Value, Perspectives and Cases on Electronic Commerce*. Wiley, NY, USA, 2002.

[SS03] A.R. Sadeghi and M. Schneider. *Digital Rights Management: Technological, Economical, Legal, and Political Aspects*, chapter Electronic Payment Systems, pages 113–137. Springer, November 2003.

[Ste97] M. Stefik. Shifting the Possible: How digital property rights challenge us to rethink digital publishing. *Berkley Technology Law Journal*, 12:137–159, 1997.

[Str03] M. Strembeck. *Engineering and Enforcement of Role-Based Access Control Policies with Context Constraints*. PhD thesis, Vienna University of Economics and BA, Department of Information Systems, New Media Lab, Austria, September 2003.

[Sun02] Sun Microsystems, Inc. Digital Rights Management: Managing the Digital Distribution Value Chain. White Paper, http://www.sun.com/, 2002.

[Sup03] Supply-Chain Council (SCC). Scor 6.0. Technical Report, http://www.supply-chain.org/, August 2003.

[SV99] C. Shapiro and Hal R. Varian. *Information Rules*. Harvard Business School Press, 1999.

[Sza02] G. Szabo. A Formal Language for Analyzing Contracts. White Paper, http://szabo.best.vwh.net/, 2002.

[TBMM01] H.S. Thompson, D. Beech, M. Maloney, and N. Mendelsohn. eXtenisble Markup Language (XML) Schema Part 1: Structures. W3 Consortium Recommendation, http://www.w3.org/TR/xmlschema-1/, May 2001.

[vWT03] B. von Walter and T.Hess. iTunes Music Store – eine innovative Dienstleistung zur Durchsetzung von Property–Rights im Internet. *Wirtschaftsinformatik*, 45:5:541–546, 2003.

[WBK03] T. Weitzel, D. Beimborn, and W. König. An Infivdual View on Cooperation Networks. In *Proceedings of the 36th Hawaii International Conference on System Sciences (HICSS), Hawaii/USA*, Januar 2003.

[Wid96] H. Widdowson. *Linguistics*. Oxford University Press, 1996.

[WIP02] WIPO - World Intellectual Property Organization. Intellectual property on the internet: A survey of issues. Technical Report, http://www.wipo.org/, December 2002.

[Wis94] Wisschenschaftskommission Wirtschaftsinformatik (WKWI). Beschluss der Wissenschaftskommission Wirtschaftsinformatik (WKWI) vom 06.10.1993. *Wirtschaftsinformatik*, 1:80 ff, October 1994.

[WL95] D. Wetherall and C.J. Lindblad. Extending Tcl for dynamic object–oriented programming. In *Proceedings of the Tcl/Tk Workshop*, July 1995.

[Wor00] World Wide Web Consortium (W3C). Document Object Model (DOM), http://www.w3.org/DOM/. W3C Technology, November 2000.

[Zhu01] K. Zhu. Internet-based distribution of digital videos: the economic impacts of digitization on the motion picture industry. *Electronic Markets*, 11 (4), December 2001.

[ZS01]   H. Zhang and E. Stroulia. Babel: Representing Business Rules in XML for Application Integration (Research Demonstration). In *Proceedings of the 23 rd International Conference on Software Engineering, Toronto, Canada*, pages 831–832. IEEE Computer Society Press, May 2001.

Index

Access control, 72
Accounting, 72
Agents, 45
Agreement category, 73
Application-specific objects, 73

Business information systems, 22
Business model, 23, 30
Business to business, 19, 45
Business to consumer, 19, 45

Combined Delivery, 43
Component framework, 104
Constraints, 132
Contract, 17, 63
Contract object, 70
Contract right, 66
Contract schema, 82, 88
Cryptolope, 46
Customer relationship management, 73

Document object model, 112
DRM, 34

E-Commerce, 17, 63
Electronic contract, 17, 63
Electronic ticket, 67

Facade, 138, 140, 157

IPR, 73

License, 35, 68

Mediator, 105, 126

Operation, 50

Pricing model, 18, 30

RDF, 20
RE generator, 46, 104, 114
RE interpreter, 46, 105, 118
REL application policies, 52, 75
REL instance, 49
REL lexis, 50
REL syntax, 50
REL vocabulary, 50
Rights data dictionary, 52
Rights enforcement, 47
Rights expression, 49
Rights locker, 45

Scenario-specific objects, 73
Secure container, 38
Secure viewer, 18, 37, 42, 103
Separate Delivery, 43
Separation of space, 51
Separation of time, 51
Service level agreement, 46, 97
Software methods, 75
Software service, 20, 75
Superdistribution, 42, 43
Supply chain, 18, 45

TCP/IP, 114
tDOM, 113, 122
Token, 37, 43

Voucher, 46, 67

Watermarks, 38, 40
Wirtschaftsinformatik, 21

XML, 20, 54
XML Parser, 56, 106
XML Validator, 56

Forschungsergebnisse der Wirtschaftsuniversität Wien

Herausgeber: Wirtschaftsuniversität Wien –
vertreten durch a.o. Univ. Prof. Dr. Barbara Sporn

Band 1 Stefan Felder: Frequenzallokation in der Telekommunikation. Ökonomische Analyse der Vergabe von Frequenzen unter besonderer Berücksichtigung der UMTS-Auktionen. 2004.

Band 2 Thomas Haller: Marketing im liberalisierten Strommarkt. Kommunikation und Produktplanung im Privatkundenmarkt. 2005.

Band 3 Alexander Stremitzer: Agency Theory: Methodology, Analysis. A Structured Approach to Writing Contracts. 2005.

Band 4 Günther Sedlacek: Analyse der Studiendauer und des Studienabbruch-Risikos. Unter Verwendung der statistischen Methoden der Ereignisanalyse. 2004.

Band 5 Monika Knassmüller: Unternehmensleitbilder im Vergleich. Sinn- und Bedeutungsrahmen deutschsprachiger Unternehmensleitbilder – Versuch einer empirischen (Re-)Konstruktion. 2005.

Band 6 Matthias Fink: Erfolgsfaktor Selbstverpflichtung bei vertrauensbasierten Kooperationen. Mit einem empirischen Befund. 2005.

Band 7 Michael Gerhard Kraft: Ökonomie zwischen Wissenschaft und Ethik. Eine dogmenthistorische Untersuchung von Léon M.E. Walras bis Milton Friedman. 2005.

Band 8 Ingrid Zechmeister: Mental Health Care Financing in the Process of Change. Challenges and Approaches for Austria. 2005.

Band 9 Sarah Meisenberger: Strukturierte Organisationen und Wissen. 2005.

Band 10 Anne-Katrin Neyer: Multinational teams in the European Commission and the European Parliament. 2005.

Band 11 Birgit Trukeschitz: Im Dienst Sozialer Dienste. Ökonomische Analyse der Beschäftigung in sozialen Dienstleistungseinrichtungen des Nonprofit Sektors. 2006

Band 12 Marcus Kölling: Interkulturelles Wissensmanagement. Deutschland Ost und West. 2006.

Band 13 Ulrich Berger: The Economics of Two-way Interconnection. 2006.

Band 14 Susanne Guth: Interoperability of DRM Systems. Exchanging and Processing XML-based Rights Expressions. 2006.

Band 15 Bernhard Klement: Ökonomische Kriterien und Anreizmechanismen für eine effiziente Förderung von industrieller Forschung und Innovation. Mit einer empirischen Quantifizierung der Hebeleffekte von F&E-Förderinstrumenten in Österreich. 2006.

www.peterlang.de

Klaus Lodigkeit

Intellectual Property Rights in Computer Programs in the USA and Germany

Frankfurt am Main, Berlin, Bern, Bruxelles, New York, Oxford, Wien, 2006.
116 pp.
ISBN 3-631-54039-6 / US-ISBN 0-8204-7734-6 · pb. € 24.50*

The author provides an overview and comparison of software protection law in two countries whose technological expertise had important influence on the digital information age – the United States and Germany. The book shows software protections under trade secrets, copyright law and patent law in the USA and Germany and also the interaction of these laws in both countries. It is a contribution to the field of comparative computer software law and will be helpful to lawyers who advise software owners and developers for the German and American markets. It is also helpful to lawyers unfamiliar with intellectual property law in general, who wish to understand the fundamental concepts of these laws for computer software.

Contents: Software protections under trade secrets · Copyright law and patent law in the USA and Germany · Interaction of these laws in both countries · Comparison of intellectual property laws of both countries

Frankfurt am Main · Berlin · Bern · Bruxelles · New York · Oxford · Wien
Distribution: Verlag Peter Lang AG
Moosstr. 1, CH-2542 Pieterlen
Telefax 00 41 (0) 32 / 376 17 27

*The €-price includes German tax rate
Prices are subject to change without notice
Homepage http://www.peterlang.de